Planet Earth
in Jeopardy

Planet Earth in Jeopardy
Environmental Consequences of Nuclear War

Lydia Dotto

Based on a SCOPE report prepared by
T. P. Ackerman, P. J. Crutzen, M. A. Harwell,
T. P. Hutchinson, M. C. MacCracken, A. B. Pittock,
C. S. Shapiro *and* R. P. Turco

Published on behalf of the
Scientific Committee on Problems of the Environment (SCOPE)
of the
International Council of Scientific Unions (ICSU)
by
JOHN WILEY & SONS
Chichester · New York · Brisbane · Toronto · Singapore

British Library Cataloguing in Publication Data:

Dotto, Lydia
    Planet earth in jeopardy: environmental
consequences of nuclear war.
    1. Nuclear warfare—Environmental aspects
    I. Title   II. International Council of Scientific
Unions. *Scientific Committee on Problems of
the Environment*
    574.5   QH545.N83

    ISBN 0 471 99836 2

Library of Congress Cataloging in Publication Data:

Dotto, Lydia.
    Planet earth in jeopardy

    1. Nuclear warfare—Environmental aspects.
1. International Council of Scientific Unions.
Scientific Committee on Problems of the Environment.
II. Title
QH545.N83D67   1986   574.5'222 ˙   85-22747
ISBN 0 471 99836 2

Printed and bound in Great Britain

Foreword

This book is based on a 2-volume technical monograph entitled *The Environmental Consequences of Nuclear War: I. Physical and Atmospheric Effects*; and *II. Ecological and Agricultural Effects*, published in 1985 by John Wiley and Sons. The principal authors were T.P. Ackerman, P.J. Crutzen, M. A. Harwell, T. P. Hutchinson, M.C. MacCracken, A. B. Pittock, C.S. Shapiro and R. P. Turco. The author of this volume, Lydia Dotto, was commissioned by an international scientific committee to write a popular account of the larger study, the objective being to reach the widest possible audience.

The Scientific Committee on Problems of the Environment [SCOPE] initiated the study in 1983 at the request of the International Council of Scientific Unions [ICSU]. A Steering Committee charged by SCOPE with responsibility for the review recruited scientists from more than 30 countries. These specialists participated through correspondence and in workshops held in London, Stockholm, New Delhi, Leningrad, Tallinn, Paris, Delft, Hiroshima and Tokyo, Toronto, Melbourne, Caracas and Colchester [UK].

The effort was made possible by the voluntary contribution of time and thought by more than 300 scientists and by financial aid from their institutions and from the Carnegie Corporation, the General Service Foundation, the Andrew W. Mellon Foundation, the Rockefeller Brothers Fund, the W. Alton Jones Foundation, the Royal Society of London, the Academy of Sciences of the USSR, the Australian Academy of Sciences, the Indian National Academy of Sciences, Maison de la Chimie [Paris], the Royal Swedish Academy of Sciences, the Royal Society of New Zealand and the United Nations University.

Recognizing the uncertainties attached to estimates of the effects of a major nuclear conflict on a scale without precedent, the Steering Committee nevertheless concludes that many of the serious global environmental effects are sufficiently probable to require widespread consideration. Any disposition to minimize or ignore those effects and the possibility of a tragedy of unprecedented dimensions would be a fundamental disservice to the future of global civilization.

In conclusion, it should be mentioned that this book is intended to provide an overview of two larger and more detailed scientific volumes and is necessarily written in a much less technical style for the benefit of lay readers. Thus, it is not possible to include all of the detailed chains of reasoning, subtleties of emphasis and shadings of meaning that are contained in the 2-volume report. However, this book is intended to reflect the overall sense of those volumes. Readers interested in pursuing the full range of evidence, questions and arguments should consult the 2-volume report. It should also be added that the authors of the technical volumes were consulted as much as was possible and their contributions are gratefully acknowledged; however, the contents of this book are the responsibility of the author, the Editor-in-Chief and the Steering Committee.

SIR FREDERICK WARNER, *University of Essex, UK* (*Chairman*)
J. BÉNARD, *Ecole Superieure de Chimie, Paris, France*
S. K. D. BERGSTROM, *Karolinska Institutet, Stockholm, Sweden*
P. J. CRUTZEN, *Max-Planck-Institut für Chemie, Mainz, FRG*
T. F. MALONE, [ICSU Representative] *St. Joseph College, USA*
M. K. G. MENON, *Planning Commission, New Delhi, India*
M. NAGAI, *United Nations University, Tokyo, Japan*
G. K. SKRYABIN, *Akademia Nauk, Moscow, USSR*
G. F. WHITE, *University of Colorado, Boulder, USA*

Contents

CHAPTER 1

The Environmental Consequences of Nuclear War

In a display at the Hiroshima Peace Memorial Museum, among half-melted lumps of metal and glass, blasted bits of brick and stone and swatches of burned clothing, sits the charred remnant of a wrist watch. The minute and hour hands have been burned away, but their shadows remain, imprinted forever by the brilliant flash of the first nuclear weapon to be exploded over a city. This image conveys a powerful message: for the wearer of the watch—and for hundreds of thousands of other residents of Hiroshima—at 8:15 a.m. on an August morning in 1945, all that they knew as a normal existence abruptly ended without warning.

The bombing of Hiroshima—and Nagasaki after it—provide our only direct experience with the consequences of nuclear weapons explosions in cities. Those events have taught us a great deal about the potential physical, biological and human impact of nuclear war. But it must be remembered that each city experienced only a single explosion of a weapon much smaller in yield than many of those stockpiled in world nuclear arsenals today. Indeed, these arsenals contain nearly 50,000 weapons with individual yields 1 to 500 times that of the Hiroshima bomb and a total explosive power about a million times greater. A single one-megaton nuclear weapon has 100,000 times the explosive power of the most powerful bomb used in World War II.

Furthermore, the environmental impact of the Hiroshima and Nagasaki bombs was geographically limited and survivors were almost immediately able to obtain medical and other assistance from outside. Thus, the impact of those bombs, though undeniably devastating for the people directly affected, was much more limited than might be expected in a full-scale war involving multiple detonations of much larger-yield weapons throughout much of the Northern Hemisphere.

The deployment of increasing numbers of more powerful weapons since the experience of Hiroshima and Nagasaki inevitably prompts the question: What would happen if many modern nuclear weapons were to be exploded? It seems obvious that the consequences would be very much more severe than they were in 1945. Perhaps even more important, however, is the fact that there could be consequences of a kind that didn't occur in Hiroshima

1

and Nagasaki and were not even contemplated until very recently—i.e., the possibility that smoke from massive nuclear-ignited urban fires could cause global-scale disruptions in the Earth's weather and climate. Recent studies have focused on these possibilities and, as a result, the following picture of the potential post-nuclear war environment has emerged:

In the aftermath of a large-scale war in which nuclear weapons were exploded in major cities, darkened skies would cover large areas of the Earth for perhaps weeks or several months, as sunlight was blocked by large, thick clouds of smoke from widespread fires. The impact would be greatest over the continents of the Northern Hemisphere, where most of the smoke would likely be produced by nuclear-ignited fires, and where average temperatures in some areas might drop some tens of degrees Celsius to below freezing for several weeks to months after the war. Climatic disturbances might persist for several years, even in countries not directly involved in the war. Rainfall in many regions of the world might be greatly reduced.

Temperature and precipitation changes could also occur in the tropics and the Southern Hemisphere—less extreme than those in the Northern Hemisphere, but still significant. Tropical and sub-tropical regions could experience unprecedented cooling and severe cold spells, accompanied by significant disturbances in precipitation patterns.

World agriculture and major ecosystems, such as forests, grasslands, and marine systems, could be severely disturbed and their plant and animal populations stressed by rapid, dramatic changes in the normal climatic regime. Crop losses, caused not only by climate disturbances but also by the post-war disruption in supplies of essential inputs such as energy, machinery, fertilizers and pesticides, could create widespread food crises in both combatant and non-combatant nations. The failure of major food production and distribution systems, and the inability of natural ecosystems to support large numbers of people, could perhaps reduce the human population of the Earth to well below current levels.

In addition to the potential climatic effects, a large-scale exchange of nuclear weapons would cause considerable devastation from the direct effects of fire, blast and local fallout of radioactivity. Other impacts could include severe disruptions of communications and power systems; reductions in the ozone layer in the upper atmosphere which protects life on Earth from the sun's biologically damaging ultraviolet radiation; intense local and long-term global fallout; and severe regional episodes of air and water pollution caused by the release of large amounts of toxic chemicals and gases.

In short, it is possible that, in the aftermath of a major nuclear war, the global environment and human social and economic systems could collapse to an extent that might preclude recovery to pre-war conditions.

This is the picture that emerges from the most recent analyses of the potential climatic and environmental consequences of a large-scale nuclear war.

These analyses strongly suggest that the indirect effects of such a war could potentially have a greater impact on human society than the direct effects. They also indicate that even non-combatant nations far removed from the actual conflict could experience significant cooling episodes and disruptions in precipitation patterns.

Of course, it is impossible to predict the consequences of nuclear war exactly and it must be emphasized that considerable uncertainties in these estimates remain [see below]. Thus the range of possible outcomes is very large; however, not all these outcomes are equally likely to occur. The extremes at either end of the scale—i.e., months of sub-freezing temperatures all over the world, or alternatively, negligible climatic effects—are much less likely than intermediate level effects, such as short freezes lasting for some days over large regions of the Northern Hemisphere and widespread continental cooling lasting for several months. The present study attempts to avoid the extremes at either end, particularly "worst case" scenarios, in favor of a middle ground that is believed to represent a more probable outcome of a nuclear war. Nevertheless, the results of these new studies support the view that *extremely serious climatic and other environmental consequences from a nuclear war are indeed possible* .

It is especially important to note, moreover, that post-war conditions *need not reach the most extreme limits of even the present estimates in order to have very serious impacts* on global agricultural and natural ecosystems and, in turn, on human society. These analyses show that many crops may be highly vulnerable to even the smaller climatic changes estimated to be possible; over the longer term, food crises could well cause many hundreds to perhaps thousands of millions of deaths among people who survive the direct effects of the war and the climatic extremes that might immediately follow. Even in the absence of extreme climatic disturbances, disruptions in social systems and global trade in food and energy resources could create famine in many countries, even those not directly affected by the nuclear exchange. Following is a summary of the new findings. In succeeding chapters, more detailed evidence is given to support these estimates of post-nuclear war climatic effects and the likely consequences for ecosystems, agricultural productivity and human populations.

SUMMARY OF CLIMATIC EFFECTS

If major cities—and particularly their fossil fuel supplies—were hit with nuclear weapons, this would almost certainly produce global climatic disturbances. It is believed that, in a large-scale war, hundreds of major urban/industrial centres in the Northern Hemisphere could be devastated by nuclear explosions, due to bombing of cities themselves or as a by-product of the targeting of nearby military and industrial installations.

The direct effects of a nuclear explosion could include ionizing radiation, blast waves and thermal radiation that could cause immediate death and destruction over an area the size of a major city. In addition, nuclear explosions could ignite urban fires unprecedented in size and intensity and massive smoke plumes could carry very large quantities of sooty smoke particles high into the atmosphere. The newly recognized atmospheric effects of this smoke are one of the major aspects of the studies on the climatic impact of nuclear war discussed in this book.

Nuclear explosions at or near the surface would also raise a large amount of dust, soil and debris, which would be drawn up in the rising nuclear fireball. Radioactive dust would contribute to local and global fallout; in addition, dust particles could be lofted high into the atmosphere and, as a result, they could also cause climatic effects. Although these effects are not estimated to be as great as those caused by smoke particles, neither would they be insignificant.

If about one-quarter of the combustible material in major Northern Hemisphere cities were to burn in nuclear-ignited fires, an estimated 50 to 150 million tonne of smoke could be produced, of which about 30 million tonne would be soot, a very black form of carbon-containing smoke that is a particularly effective absorber of the sun's energy. The black smoke produced by the flaming combustion of fossil fuels [e.g. oil, gasoline, kerosene] and their derivatives [e.g. plastics, rubber, asphalt] would contain the highest fraction of soot.

Fires ignited in forests and wildlands would also produce large amounts of smoke but, since this smoke would contain a smaller fraction of soot, the contribution to climatic effects would probably be secondary to that caused by the smoke from urban fires.

Because they are characterized by strong updrafts, the plumes rising from intense surface fires could loft smoke particles to altitudes as high as 10–15 kilometers, into the upper *troposphere* or lower *stratosphere*.[1] The number of smoke particles carried aloft would depend on the rate at which smoke was generated, the intensity of the fires, local weather conditions, and whether the particles were removed by rain on their way into the upper atmosphere.

In many cases, the smoke plumes could enter towering thunderclouds triggered by the intense heat from the fires. Such clouds engulfed the plumes from the Hiroshima and Nagasaki bombs, producing a "black rain" containing soot, dust and ash that fell for several hours after the explosion.

---

[1] The *troposphere* is about the lowest 10–15 kilometers of the atmosphere, characterized by a decrease in temperature with altitude. It contains 90% of the mass of the atmosphere and most of what is known as "weather." The *stratosphere* is located above the troposphere, extending to about 50 kilometers; there, the temperature is either constant or it increases with height. The stratosphere contains the ozone layer that protects the earth's surface from the sun's ultraviolet radiation. The boundary between the troposphere and the stratosphere is called the *tropopause*.

This demonstrates that some particles are promptly rained out, but removal processes are not yet well understood and it is uncertain what fraction of the particles originally lofted in the rising plumes would be rained out very quickly. In most current studies, it is assumed that 30% to 50% of the particles would be removed by precipitation within a few hours or days.

As mentioned earlier, sooty smoke is a strong absorber of solar energy. Dense clouds of smoke particles floating high in the Earth's atmosphere could intercept a substantial fraction of the sun's incoming radiation, preventing it from reaching the surface. This would cause heating in the upper atmosphere and cooling at the surface. If 30 million tonnes of soot particles were spread over the mid-latitudes of the Northern Hemisphere, solar radiation at the ground would be reduced by at least 90%. During the first few days, the smoke particles could be concentrated in very dense patches that could be carried long distances by the winds. This could cause quite variable conditions at the surface, including brief episodes of severe cooling. Under the dense smoke patches, light levels could be reduced to nearly zero; even after the smoke had spread widely, the levels might be only a few percent of normal, on a daily average.

There is no evidence that meteorological processes would remove significant quantities of smoke particles within the first few days after the nuclear explosions. After a few days, the upper troposphere would stabilize and solar heating of the smoke-laden clouds could cause the smoke particles to be lofted even higher into the atmosphere, above the region where precipitation might wash them out.

Within a few days, continental-sized smoke clouds could spread over North America, Europe and much of Asia. The potential climatic impact of these clouds has been estimated using a variety of climate computer models. All simulations strongly indicate that smoke from fires could cause large, global-scale disruptions in normal weather patterns. These models still have many uncertainties and simplifications; for example, they are inadequate in projecting the removal of particles from the atmosphere. However, they have been substantially improved recently to include new features that are important in considering the post-war climatic effects of large quantities of smoke.

The estimates derived from computer simulations are summarized below. They assume that large quantities of smoke would be carried up into the atmosphere at altitudes of several kilometers or more in the Northern Hemisphere. It will be apparent that the season in which the war occurred would be a critical factor influencing the subsequent climatic impact.

- If the smoke injection were to occur between the late spring and early fall in the Northern Hemisphere, average land surface temperatures beneath dense smoke patches in continental interiors could decrease by 20–40°C below normal within several days. Some of the smoke clouds might be carried long distances and cause cooling episodes as they passed, possibly causing sudden freezes. Weather conditions could be quite variable during this time if periods of dense smoke, during which virtually no sunlight would reach the surface, alternated with periods of thin or no smoke, during which substantial fractions of the normal amount of sunlight would reach the surface.

• Smoke particles would be spread throughout the Northern Hemisphere, although somewhat unevenly, in 1 to 2 weeks. For smoke injections in late spring to early fall, the absorption of solar energy by the particles would cause a rapid warming of the air in the upper atmosphere, which could loft the particles even higher, where they could remain for months to years.

• Average summertime land surface temperatures in the Northern Hemisphere mid-latitudes could drop to levels typical of early winter for periods of weeks or more and precipitation could be essentially eliminated. Fog and drizzle might occur, especially in coastal regions. Periods of very cold [mid-winter-like] conditions might occur in continental interiors.

• If the smoke injection occurred in winter, light levels would be greatly reduced but the initial temperature and precipitation disturbances would not be as severe as those which might occur during summer. In fact, in many locations, conditions might be indistinguishable from an unusually cold winter. However, such conditions would occur simultaneously over the entire mid-latitude region of the Northern Hemisphere. More important, freezing cold air outbreaks might invade southerly regions that rarely or never normally experience frost.

• For large smoke injections, temperatures in any season in sub-tropical latitudes of the Northern Hemisphere could drop well below typical cool season conditions. In areas not strongly influenced by the warming influence of the ocean, temperatures could be near or below freezing. The monsoons, a critical source of precipitation in sub-tropical Asia and Africa, might fail, although precipitation could increase in some coastal areas.

• The strong solar heating of the smoke injected in the Northern Hemisphere beween spring and autumn would carry it upward and toward the equator. Witnin 1–2 weeks, thin smoke clouds might appear in the Southern Hemisphere, followed by a thin, more uniform veil of dark smoke. Modest cooling could occur in regions not influenced by the warming influence of the ocean; however, since this would occur in the Southern Hemisphere cool season, these initial temperature reductions would not likely exceed 10–15°C degrees.

- There are still many uncertainties in estimating the recovery of the atmosphere from a nuclear war. A period of acute climatic disturbance could last for several weeks or months. Present estimates are that smoke initially injected at 10 kilometers or higher could remain in the atmosphere for a year or more and could cause global-scale cooling of several degrees and continued reductions in precipitation.

OTHER EFFECTS OF NUCLEAR WEAPONS EXPLOSIONS

- *Nitrogen oxides* created by nuclear fireballs and carried up into the stratosphere could deplete the ozone layer that protects life on Earth from the sun's biologically damaging ultraviolet radiation [UV-B]. It is estimated that Northern Hemisphere stratospheric ozone could be reduced by as much as 20% to 30% within six months to a year. It is possible that "ozone holes", with reductions of perhaps 70%, could occur briefly. Under a clear atmosphere, ozone depletion would cause UV-B levels at the surface to increase, but UV-B would be blocked for a few weeks to months by smoke in the atmosphere. However, the smoke might also play a role in destroying ozone directly and in changing air temperatures and circulations patterns. It is possible that the ozone effects could be larger and longer-lasting.

- Large amounts of *chemical pollutants* could be released into the atmosphere after a nuclear war, including carbon monoxide, hydrocarbons, nitrogen oxides, sulphur oxides, hydrochloric acid, heavy metals and a variety of other toxic chemicals. These could be directly or indirectly harmful to humans and many other organisms for periods of hours to years. Depending on a variety of factors, the acidity of rain could be increased by 10 times [a decrease of one pH unit] over current levels in polluted industrial areas for a month or more and cold acid fogs may form near the ground. Strong temperature inversions might trap toxic compounds from smouldering combustion in pockets near the ground in high population and industrial regions, causing significant health risks.

- During the first few days after a major nuclear exchange, heavy local *fallout* would occur within the vicinity of explosions, especially downwind of missile silos and other "hardened" targets receiving surface bursts. Vast land areas of combatant countries would receive lethal doses from external gamma-radiation alone. Over longer periods, assuming no sheltering, more widespread but less intense global fallout would occur, leading to an increase in the incidence of cancers and other health hazards. Additional doses of radiation could be received from several sources, including lower-yield weapons and ingestion and inhalation of radioactive particles in food, water and air. Survivors would be more vulnerable than normal to radiation effects because of other injuries and environmental stresses and because of the probable lack of medical help.

It is possible, though considered unlikely by some analysts, that Northern Hemisphere doses could be increased by a few times if the civilian nuclear power system and military reactors were directly hit with nuclear weapons. In the Southern Hemisphere, local fallout would be important only within a few hundred kilometers downwind of any surface bursts and global fallout doses should be relatively insignificant.

• If nuclear weapons are exploded high in the atmosphere, large areas of the Earth might be subjected to electromagnetic pulse [EMP], which can upset or destroy communications, power and electronic systems. The loss of these systems at the outset of a war could compound the confusion and chaos at a time when critical decisions are being made about the use of nuclear weapons.

SUMMARY OF BIOLOGICAL FINDINGS

The approach taken in evaluating the biological consequences of nuclear war was to assess the *vulnerability* of natural ecosystems and agricultural systems to the range of climatic and other disturbances estimated to be possible after a full-scale exchange of nuclear weapons.

This analysis resulted in the following major conclusions:

• Both natural ecosystems and agricultural systems are extremely vulnerable to the climatic and other stresses that could result from a major nuclear war. For perhaps thousands of millions of people not subjected to the direct effects of the war, the major post-war risk could be starvation resulting from the failure of crops, the loss of international trade in food and energy supplies and the depletion of food stores.

• Substantial reductions in crop yields and even widespread crop failures are possible in response to stresses caused by post-war reductions in temperatures, precipitation and light levels. For example, temperature reductions could result in a shortening of the growing season simultaneously with an increase in the time needed for crops to mature; the combination of these two factors could result in insufficient time for crop maturation before the onset of killing cold temperatures. The occurrence of brief episodes of chilling or freezing temperatures at critical times within the growing season would also cause large crop losses.

• Even in the absence of climatic disturbances, the global agricultural production and food distribution system is highly vulnerable to disruption after a nuclear war, because of the potential loss of international trade; food transportation, storage and distribution facilities; agricultural machinery; energy supplies; fertilizers and pesticides; seed supplies and other subsidies to agriculture provided by humans.

- In response to a combination of climatic and other factors, it is possible that food production in most of the Northern Hemisphere and much of the Southern Hemisphere could be virtually eliminated for at least a year after a large scale war.

- Food exports probably would also cease because of the general disruption of world trade and because the major Northern Hemisphere grain-exporting countries are among the likely combatants. In the absence of agricultural production and food imports, the amount of food held in storage would become the critical factor controlling human survival in the first months after the war. *An analysis of world food stores indicates that, for much of the world's population, stores would be depleted before agricultural productivity could be resumed. As a result, the majority of people on Earth face the risk of starvation in the aftermath of a large-scale nuclear war; for those in non-combatant nations, famine could be the major consequence.*

- The stresses placed on the agricultural system might result in greater demands being placed on natural ecosystems for food. However, even in the absence of post-war climatic changes, natural ecosystems could support only a very small fraction of the current global population. Under optimal conditions, they could not replace agricultural systems as a source of sustenance for humans.

- Natural ecosystems would, in any event, be highly vulnerable to the climate disturbances estimated to be possible after a nuclear war. The responses of ecosystems—ranging from arctic tundra to tropical rainforests and from deserts to ocean and freshwater ecosystems—would be highly variable. In general, temperature effects would dominate in land ecosystems in the Northern Hemisphere; reductions in sunlight would have the most impact on oceanic ecosystems; and precipitation changes would be most important for grasslands and Southern Hemisphere ecosystems. It is possible that the combination of these climatic factors could have a much greater impact than an analysis of their individual effects might suggest.

- Biological systems are so sensitive to even relatively small changes in climatic conditions that devastating consequences could result from even the smaller climatic effects projected to be possible in the post-war environment. *Thus, resolving some of the remaining uncertainties about the physical effects of a major nuclear exchange is not absolutely necessary in assessing the human impact of food shortages, since even the lower estimates of physical effects could be devastating to human populations on a large scale.*

• Longer-term climatic disturbances, if they were to occur, could be at least as important to human survival as the acute-period extremes of light and temperature reduction. Resolving uncertainties about reductions in precipitation is particularly important.

• Global fallout is not likely to result in major ecological, agricultural or human effects. Local fallout, however, could have a serious impact on combatant and adjacent countries. Various estimates suggest that millions to perhaps hundreds of millions of people could die from the effects of lethal local fallout and it is likely that these direct effects would far exceed the indirect effects resulting from radiation damage to ecosystems. Agricultural and natural ecosystems would, however, be affected; large areas in the mid-latitudes of the Nothern Hemisphere, including freshwater and marine ecosystems, could be radioactively contaminated for long periods of time. Radiation can be concentrated or "magnified" by biological systems, so that organisms at the higher levels of a food chain, including humans, could receive doses increased by factors of several thousand from eating contaminated food. It is unlikely that an efficient means of decontamination would be available in the post-nuclear war world.

The studies discussed in succeeding chapters do not explore the social and psychologial effects of a major nuclear war, particularly the extent to which human behavior and social organization might change in the aftermath of the war. No society has ever been subjected to the kind and magnitude of environmental stresses postulated in these scenarios, so trying to predict social reponse on the basis of historical catastrophes would be, at best, highly speculative. It seems obvious that social and psychological factors would play an enormous role in shaping the nature of the world that would emerge. This is a subject that deserves investigation by social scientists.

UNCERTAINTIES

The reader will notice frequent cautions in the following chapters that some findings cannot be taken as reliable predictions because of the uncertainties that remain. The problem of scientific uncertainty is a difficult and frustrating one for many non-scientists—who often tend to look to the scientific community for the "right" answers—but uncertainty cannot be avoided when large, complex systems like the Earth's atmosphere and oceans, agricultural systems and natural ecosystems and, not least, human society, are simultaneously involved in the analysis.

The uncertainties associated with the studies cited are of two basic kinds:

• Those resulting from human actions, e.g. assumptions about the nature of a nuclear war, including factors such as the numbers of weapons exploded; yields; targets; time of day and season and height of bursts.

• Those resulting from the incomplete state of knowledge concerning physical and biological processes and the inability to simulate them precisely in mathematical computer models.

The first sort of uncertainty is very difficult to reduce; clearly, the exact circumstances of a large-scale nuclear war are impossible to specify. Thus, although it is necessary to make some assumptions regarding the nature of the war, detailed war scenarios must be regarded as speculative. One approach, taken in many of the studies cited in this book, is to base the analysis of environmental effects on a plausible circumstance or range of circumstances—for example, the injection of a given mass of smoke into the atmosphere. In essence, the approach is to say: If "X" occurs, then "Y" could be the consequence.

Uncertainties associated with the state of scientific knowledge often can be reduced with further research. However, short of actually doing the "experiment"—i.e., starting the war—many of the details will continue to be in doubt.

Two important points about uncertainty must be kept in mind. First, *uncertainties go in both directions*. If and when they are resolved, the outcome could be less serious than originally thought or it could be more serious. It follows that the *presence of uncertainty does not mean that there is no problem*. Uncertainty implies that the outcome cannot be predicted with complete confidence; it does not necessarily imply that the outcome won't be serious. Because of the problem of uncertainty, the analysis presented in this book attempts to avoid the extremes in favor of a middle ground.

COMPUTER MODELLING

Computer models attempt to simulate the real world. In general, they consist of a complex series of mathematical equations that represent the system being modelled—for example, the Earth's atmosphere or oceans—as accurately as possible, given the current state of scientific knowledge. Models can be run forward in the computer to predict how the system might change over time; this is how weather forecasts are made. They can also be used to investigate what might happen if various stresses or changes were to be introduced into the system—for example, the injection of large amounts of soot into the upper atmosphere in the aftermath of a major nuclear war.

The studies cited in this book employed a variety of computer models in two general classes—climate and biological models. These, too, are subject to many uncertainties, stemming mainly from simplifying assumptions and an incomplete understanding of many important physical and biological processes. Although they are intended to examine conditions that have not yet happened in the real world—indeed, in the case of nuclear war, conditions that we hope will *never* happen—they nevertheless depend on inputs of accurate information about the real world. Large natural systems such as the Earth's atmosphere, oceans and ecological and agricultural systems are very complex and the scientific knowledge of many of the important processes that occur within them is lacking or incomplete.

However, it can be said that much research has been done on this problem in recent years and there have been substantial improvements in many of the computer models used. Comparing the models with real-world conditions has demonstrated that many have very good predictive value. Several important advances have recently been made in the climate models used for studying the potential consequences of a large-scale nuclear war, and they now include some important features which earlier modelling studies were criticized for neglecting. The details of these and other modelling studies are discussed at appropriate points throughout the book. The following discussion focuses on more general features.

Climate Models

These models are typically classed as *1-dimensional* [*1-D*], *2-dimensional* [*2-D*], and *3-dimensional* [*3-D*]. A 1-D model includes details of the vertical structure [up and down], but not the horizontal. A 2-D model includes both vertical and horizontal variations [usually in the north-south direction only] and a 3-D model include up and down, east-west, and north-south variations. Each of these types of models may simulate how the system changes over time.

Different models [or combinations of models] are suitable for studying different problems. While 3-D models are more comprehensive in their treatment of spatial variations and atmospheric motions, they consume much more computer time and human resources than 1-D and 2-D models and therefore have to leave out more of the physical and chemical processes that can often be included in the 1-D and 2-D models. In studying any given climate problem, there are complex trade-offs among the strengths and weaknesses of each type of model—how much detail is needed, how many dimensions, what computer resources are available and the amount of computer time [and therefore the cost] involved. For many problems, 1-D and 2-D models serve very well. For very complex problems, researchers will often use all three kinds of models to study different aspects.

The more comprehensive 3-D models available today are able to reproduce many general features of the Earth's climate very well—for example, the seasonal changes in temperature and winds or the variations in surface temperature between day and night. In fact, with appropriate modifications, these models can even reproduce the dramatically different climatic conditions on other planets such as Venus and Mars.

Biological Models

Several agricultural and ecological models were used in this study. Agricultural models can, for example, be applied to forecasting crop yields. Real-world information on important factors, such as temperatures, light levels, precipitation and nutrients in the soil, is put into these models. Physiological data on plant responses, obtained in laboratory and field studies, are also included—for example, the temperature regime and length of growing season required for a given crop to mature.

Ecological models are designed to describe the biological processes occurring in ecosystems and the inter-relationships of the plant and animal species within them. These relationships are based on the flow of energy through the food chain: primary producers are plants that produce organic material directly from the sun's energy through photosynthesis; this organic material provides the food [energy] for the higher levels of the food chain, e.g. plant-eating animals [herbivores] and meat-eating animals [carnivores]. Other ecological models are more limited in scope—e.g. models that deal only with tree productivity in forests—but incorporate other factors, such as competition among individual organisms for resources.

In the following chapters, the role of these climatic and biological modelling studies in projecting the potential environmental consequences of a nuclear war is discussed in greater detail.

CHAPTER 2
The Nuclear Exchange

The exact nature of a large-scale nuclear war would be determined by human decisions made in the context of unique global political circumstances—factors which are, and will remain, largely unpredictable. Thus it is impossible to say what such a war would be like—how or where it would start, how many weapons would be used, where they would explode, how much devastation they would cause. And yet, any attempt to understand what the world might be like after a nuclear war must necessarily start with some assumptions about the nature of the war itself. Perhaps the most logical place to start is with an examination of the existing global stockpile of nuclear weapons.

World Arsenals

Nuclear weapons come in different yields and are designed to operate over different ranges. For the purposes of this discussion, they can be divided into two basic categories. Tactical weapons are short-range, generally low-yield devices intended for use on the battlefield. Other types of weapons,[1] including the highest-yield devices, are intended for use over longer ranges, from medium or intermediate distances to long-range or intercontinental distances.

Although the actual inventories of nuclear weapons are secret, authoritative unclassified estimates of existing and projected weapons systems indicate that world arsenals amount to about 50,000 weapons, with a combined explosive yield of nearly 12,000 megatons.[2] Most of these weapons are approximately evenly divided between the United States and the USSR, with a smaller number in Great Britain, France and the People's Republic of China.

---

[1] The terms "strategic" and "theatre" are avoided in this discussion because there is a lack of consensus internationally about the definitions of these terms and the exact dividing lines regarding the ranges of different types of weapons.

[2] A megaton is the amount of energy that would be released by the simultaneous explosion of roughly one million tonne of TNT. This is about 100,000 times the explosive power of the largest bomb [the "blockbuster"] used in World War II. The global inventory of 12,000 Mt is equivalent in explosive energy to nearly one million Hiroshima bombs. Over the past two decades, there has been a trend toward stockpiling lower-yield bombs in nuclear arsenals; however, this trend seems to have halted or reversed with the latest systems.

Of the total, tactical weapons account for about 25,000 devices, amounting to several hundred megatons.

These estimates, made in 1984, are likely already outdated, but they are probably not off by more than 35%. Dramatic changes in the arsenals are not expected during this decade, and probably the next, under existing programs and treaties, so these estimates provide a reasonable starting point for discussing what a global nuclear war might be like, even allowing for any reduction agreed at the Geneva talks begun in 1985.[3]

Targets and Targeting Strategy

The number of potential target sites for nuclear weapons is estimated to be about 40,000 or more for each of the two major superpowers. Although the specifics are secret, it is possible to deduce a general order of priority. The highest targeting priority undoubtedly goes to key military facilities and forces, including intercontinental ballistic missile [ICBM] silos and command centres; air and submarine bases; weapons production and storage facilities; *command*, control, communications and intelligence [C^3I] centres; and submarines at sea. Certain civilian facilities would also be regarded as militarily important and are probably included on nuclear target lists: major air fields, fuel depots, transportation and storage facilities, and industries producing essential goods such as petroleum, lubricants, electrical power, steel and chemicals. [Nations such as Japan, Middle-Eastern countries, Australia and South Africa might be targeted in a military campaign to deny their use as staging areas and forward bases or as suppliers of manufactured and raw materials.]

Major cities are also likely to be affected. Even if they were not intended to be hit directly—as some policy-makers maintain—they could still suffer massive damage because of their proximity to high-priority military and industrial targets. Many military bases, ports, airfields and transportation and C^3I facilities are located in or near large cities, as are many major industries that would support a war effort. Since an average strategic weapon can damage up to several hundred square kilometers, extensive devastation of cities is to be expected in any sizeable exchange of nuclear weapons. It is also possible that there would be deliberate strikes against cities to cripple attempts at recovery and rebuilding.

A strategic nuclear attack intended to destroy the military power of the adversary is called a "*counterforce*" *attack*. One intended to devastate the adversary's economic and social base and slow down recovery by destroying

---

[3] This study does not consider the potential role of strategic defense systems, including the space-based system known as "Star Wars", because the feasibility of such systems has not been demonstrated. They would not, in any case, be put in place for decades and their effect on the numbers of ground-based nuclear weapons is unknown.

cities and industries is called a *"countervalue" attack*. The latter represents the ultimate cost of a global nuclear war. The blasting and burning of cities and industries, either directly or as a by-product of attacks on military installations, would not only kill millions of civilians and destroy the social and economic fabric of society, it would generate most of the smoke that could trigger severe global weather effects.

Escalation

A massive strategic attack "out of the blue", without prior crisis or conflict, is possible but not considered very probable. A nuclear war would more likely start with conventional hostilities, perhaps on a limited level, and escalate to an all-out nuclear exchange. In a major crisis, all nuclear forces would be put on alert, thus greatly increasing the magnitude and speed of the ultimate exchange. It is considered unlikely that a global nuclear war would be triggered by accident or by an isolated act of terrorism. However, some military planners seem to take seriously the option of launching a massive pre-emptive strike in the midst of a deepening political/military crisis that appears to be leading to full-scale war; this would tend to increase the danger of escalation.

There is debate on the matter of escalation. Some strategists contend that a nuclear exchange could be limited or controlled, or that it might be automatically self-limiting. Others, however, believe that this is not possible—that, with present and projected forces, a nuclear war, once started, would inevitably become uncontrollable. Since the matter is essentially impossible to forecast, it is wise to consider the effects of a large-scale exchange of weapons which is plausible given the current stockpiles in world arsenals.

SCENARIOS FOR A NUCLEAR WAR

An extremely complex chain of events must be evaluated in order to arrive at any conclusions about the likely environmental impact of a nuclear war. The first links in the chain concern the nature of the war itself—the numbers and yields of weapons exploded, the numbers and kinds of targets hit, and whether the weapons explode in the air or at the surface. The significance of these factors with respect to the environmental aftermath of the war will be discussed in more detail in this and subsequent chapters. However, as was pointed out in Chapter 1, there are many uncertainties associated with these factors which are largely unresolvable because of their dependence on human decision-making and on the performance of systems never tested under wartime conditions. There are, in fact, far too many possible nuclear war scenarios to evaluate; so it is necessary for the purposes of analysis to make assumptions and create scenarios for a nuclear war based on the best available knowledge of existing nuclear arsenals and strategic thinking. These "example" nuclear wars are not intended as forecasts—it is impossible to predict whether these, or any other scenarios, will actually occur. Instead, they are intended as an exercise in "what if?"—an effort to examine the possible or probable environmental consequences should anything approximating the scenarios actually happen.

While the course of a nuclear war may be unpredictable, it is worth asking whether a massive nuclear exchange of thousands of megatons is even credible. The present study concludes that it is, since the weapons for conducting such a war, and elaborate plans of action, exist. *The fact is that large numbers of weapons have been deployed and the intention to use them in various circumstances has been seriously expressed. This implies that a very large-scale exchange of weapons is indeed credible.*

Scenarios

Since the mid-1970s, more than half a dozen major studies have examined the potential environmental consequences of a major nuclear war. Each of these studies has had to make some basic assumptions about the nuclear exchange itself; the ones that focus specifically on the atmospheric effects made assumptions about the factors critical to analyzing climatic consequences, which include, in addition to total yield exploded and weapon sizes, the megatonnage exploded in industrial/urban centers and whether the explosions were *ground bursts* [explosions that occur at or near the surface] or *air bursts* [explosions that occur within several kilometers of the surface and in which the fireball does not contact the ground.]

In most of these studies, it was assumed that no more than half of the total world nuclear arsenals would be detonated. In fact, some of the adopted scenarios assumed the exchange of only about one-quarter to one-fifth or less of the world's inventory of weapons.

A hypothetical strategic nuclear war may be divided into four phases. It is assumed the war could escalate from one phase to the next over time. Although it is possible to contemplate smaller nuclear exchanges, and it is also possible that the conflict could end at each stage, it is not unreasonable to remain skeptical of the concept of controlled or limited nuclear warfare. The four hypothetical phases [not including a possible tactical "triggering" phase] are:

[1] an initial counterforce attack and response against military targets;

[2] extended counterforce attacks against secondary military bases, with resultant damage to cities;

[3] massive strikes against the industries that support military operations, along with continuing military strikes;

[4] the final stage, including countervalue attacks against the economic infrastructure, intended to retard postwar recovery, which would result in extensive damage to cities.

The following table shows what has been assumed in this study, regarding weapons yields in megatons distributed between military and industrial/urban targets for the four phases outlined above.

Phase	Yield|Mt	Number of Warheads	Military yield [Mt]		Industrial/Urban yield [Mt]	
			Air	Surface	Air	Surface
[1]	2000	5000	1000	1000	0	0
[2]	2000	3800	750	750	250	250
[3]	1000	1200	250	250	500	0
[4]	1000	2600	250	250	500	0
Total:	6000	12600	2250	2250	1250	250

The destructive power released in each stage is enormous compared with any previous war, but in fact, the total of 6000 megatons assumed for the cumulative total of the four stages represents about half of the total global inventory. The other half is destroyed by accurate targeting. This scenario also makes certain assumptions about the yields of individual weapons—a factor that has an important bearing on some of the environmental consequences examined in later chapters.

The scenario distinguishes between ground bursts and air bursts.[4] There are significant differences between the two types with respect to subsequent environmental effects. Ground bursts throw large quantities of radioactive dust into the atmosphere, which, in addition to causing climatic effects that are of primary concern here, produce lethal plumes of radioactive fallout over large areas. Compared with ground bursts, air bursts produce little fallout, but larger areas of blast and fire damage for the same yield and thus potentially more smoke. Ground bursts would be used against military targets that have been "hardened" [specially reinforced] to withstand blast effects; in countervalue attacks against cities, air bursts would probably predominate.

The present scenario includes the entire range of targets previously discussed—military, industrial and urban. It does not envision an attack without warning; it assumes that the conflict escalates over time and that

---

[4] It is assumed that only a small percentage of weapons would be detonated high in the atmosphere or on the ocean surface. The high atmospheric explosions would create an electromagnetic pulse which could cause massive disruption of communications and control [see Chapter 3] and therefore might have an effect disproportionate to their actual numbers if they encouraged escalation.

all major military forces can be put on alert as the crisis deepens. It also assumes that substantial military forces and weapons are destroyed early in the conflict.

The scenario includes potential damage to cities caused by strikes against nearby military targets. In the initial counterforce phase, none of the explosions is assumed to cause industrial/urban damage but during the escalating stages, such damage is assumed to be caused, either directly or as a by-product of military strikes, by 25% of the explosions in the extended counterforce phase and by 50% of the explosions in each of the two final phases. The explosions in urban areas, primarily airbursts in this scenario, are particularly important in evaluating environmental effects because they would ignite highly combustible, soot-generating materials accumulated in cities, particularly fossil fuel products and their derivatives such as plastics, rubber and asphalt.

Although the scenario does not include the use of tactical weapons, these might also be important, since they could not only produce extensive fires and radioactive fallout, but could well act as a trigger for a much larger exchange.

This description is intended only to indicate that a large-scale nuclear exchange could occur, given existing weapons and strategies. Subsequent chapters examine potential environmental consequences for a broader range of possibilities not tied directly to this one scenario. But the study does proceed on the assumption that a massive exchange of nuclear weapons is indeed plausible. And it adds one cautionary note: there is cause to believe that the number of explosions needed to create severe atmospheric effects may be much smaller than envisaged in this scenario, and could still have major environmental consequences.

CHAPTER 3
Fire, Blast and Other Immediate Effects

Within a millionth of a second after a nuclear weapon is detonated, enough energy is released to heat the surrounding air to tens of millions of degrees Celsius, forming a hot, buoyant bubble of gases, or *fireball*. The fireball rises and expands rapidly and a high-pressure shock wave moves out from it. There is also a very intense "thermal pulse" consisting of energy similar in wavelength to sunlight, which appears as a brilliant flash of light lasting for a few seconds.

As the fireball rises, it spreads out to form the distinctive mushroom cloud. The ascent of the fireball creates strong suction at ground level. This updraft, containing dust, smoke and radioactivity created by the explosion, forms the "stem" of the mushroom cloud. The fireball cools as it rises and expands and, in the case of explosions in the megaton range, the top of the mushroom cloud stabilizes within the stratosphere.

For weapons exploded between the surface and an altitude of 30 kilometers, about one third of the energy is released in the thermal pulse, half as blast and shock, and the remainder as nuclear radiation.

FIRE AND BLAST

Fires

Like strongly focused sunlight, the intense thermal pulse of a nuclear fireball can ignite fires over a large area, depending on the yield of the weapon and the nature of the materials available for burning. A one-megaton air burst can ignite light materials, such as paper, dry leaves and some fabrics at distances of up to 15 kilometers, and heavier materials, such as wood and combustible roofing, at distances of a few kilometers. Under typical atmospheric conditions, such a burst over flat terrain could ignite fires in dry flammable materials in an area of 200 to 600 square kilometers.

All types of urban areas—city centres and suburbs, commercial and industrial areas—would be subject to burning after a nuclear explosion. However, the exact extent and intensity of the fires could vary considerably from location to location, depending on such factors as the type and amount of fuel available for burning and its distribution, the numbers, yields and locations of weapons exploded, topography, and the local wind speed and direction.

Near the central point of the explosion, even buildings of heavy construction would collapse; non-flammable rubble of concrete and steel would cover some of the flammable materials, but these rubble layers would probably account for less than 10% of the total potential fire area. Within this zone, the thermal radiation would be so intense that many materials might be simultaneously ignited in fires which would continue to burn in the rubble.

Intense city-sized fires would start that could create severe local weather, including thunder, lightning and violent winds, lasting for hours after the explosions. It is difficult to estimate how fires would develop and spread in cities after a nuclear explosion.

Under certain circumstances, it is possible that a *firestorm* might develop within the first few hours after the explosions. This is an unusually intense stationary fire that produces very strong updrafts. Temperatures at the surface in the middle of the firestorm could become hot enough to incinerate all flammable materials and melt glass and some metals. The Hamburg firestorm in World War II reduced all combustible materials to ashes.

A firestorm also creates strong inflowing surface winds as air is sucked in from the periphery. Dust and ash created by the explosions and fires can be whipped into violent whirlwinds. The powerful updrafts can also be an efficient means of lofting smoke particles high into the atmosphere.

If a firestorm did not develop soon after the explosions, a conflagration— a spreading fire—might occur instead, moving out from the central blast-damaged area, engulfing new areas in fire and leaving large burnt-out areas in their wake. The peak fire intensity would likely occur within several hours after the explosion and, within the first day, the fires could spread well beyond the original ignition zone. The potential burn-out urban areas could be as large as 1000 square kilometers for a one-megaton explosion.

Forests and wildlands might also be set on fire by a nuclear explosion, although their susceptibility varies greatly according to season [e.g. fewer fires would start in winter]. The thermal pulse might not only ignite dry litter, it might have a drying effect on moist materials, including live vegetation, making them more likely to burn. The explosion might also add to the amount of fuel available for burning by knocking down foliage and branches into the fire zone. These effects can only be estimated from natural and man-made fires, but it is possible that wildland fires might be more likely to start, consume more fuel and be more intense than forest fires.

Blast

For low-altitude nuclear explosions, about half of the energy is carried away in the shock wave. The destructive power of the shock wave is measured in terms of *overpressure*, given in pounds per square inch [psi].[1] Overpressures experienced at any given distance vary according to the yield of the weapon, topography and shielding by intervening structures and the height of burst. Severe winds are also created by the nuclear blast; they are quite destructive to wind-sensitive structures, such as radar dishes, some buildings and trees.

For low-altitude air bursts, the blast wave is reflected from the ground, adding to and effectively doubling the energy of the original wave. The air shock wave causes ground shock when it slaps the surface; however, the strongest ground shocks are created by explosions at or near the surface or underground. It is possible that underground explosions could trigger earth movement along geological faults or landslides, but these effects would be highly dependent on the specific geological conditions at the site of the explosion.

All structures are susceptible to damage from blast. Typical residential housing can be badly damaged at 2 psi overpressure and crushed at 5 psi. Concrete and steel buildings can be destroyed by 10–15 psi, although interiors and facades can be destroyed at much lower overpressure levels. Glass windows can be shattered at 1 psi. Aircraft are damaged at 1–3 psi and liquid storage tanks can be split apart at 3–10 psi. One of the major hazards resulting from blast is flying debris; for example, while the human body can sustain 10 psi static overpressure before serious injury occurs, severe wounds can be caused by flying glass and other debris at 1–2 psi.

Fire/Blast Interaction

Blast damage is likely to have the net effect of increasing fire damage. Although the blast winds can blow out some primary fires started by the nuclear thermal pulse, some of these fires would be rekindled from smouldering materials. More significantly, the devastation caused by blast—broken gas lines, electrical short circuits, etc.—would result in secondary fires, which, based on the experience in Hiroshima and Nagasaki, might well be as important as the primary fires. The destruction of doors, windows and firewalls and the spilling of flammable liquids, fuels and petrochemicals would also contribute greatly to fire spread. In the general devastation and chaos, which would likely include the loss of water pressure, effective fire-fighting would be virtually impossible, even assuming there would be enough uninjured people to do the fire-fighting. This was the situation that occurred in Hiroshima and Nagasaki.

---

[1] The overpressure of a shock wave is the sudden increase in air pressure over local atmospheric pressure just behind the shock front. Atmospheric pressure varies with altitude; at sea level it is 14.7 psi. The sonic boom of an aircraft travelling faster than the speed of sound is an example of a weak shock wave.

OTHER IMMEDIATE EFFECTS

Nuclear Radiation

Within a fraction of a second after a nuclear fission explosion, about 5% of the energy is liberated as *prompt nuclear radiation*. This radiation can have important effects on humans only within a few kilometers of the explosions. Longer-term and more widespread effects are created by delayed *nuclear radiation* from fallout, which accounts for about 10% of the total energy yield of a fission explosion. Dust and debris picked up by the fireball are contaminated with radioactive elements. In a ground burst, the radioactivity is mainly deposited on the dust particles sucked up by the fireball; in an air burst, the radioactivity is mainly carried by particles formed from bomb debris—metals and other materials from the bomb itself.

In ground bursts, about 50% of the total radioactivity produced by the explosion falls out on the heaviest particles within the first 24 hours. This is called *early* or *local fallout*.The other 50% and virtually all of the radioactivity in an air burst remains in the atmosphere for longer periods on fine particles which are widely dispersed before falling to the surface. This *global fallout* is less intense than local fallout, but it is more widespread. A more detailed examination of radiation and fallout is contained in Chapter 7.

Electromagnetic Pulse [EMP]

If a nuclear weapon were exploded high in the atmosphere [above about 40 km], thermal radiation and blast effects at ground level would be minimal. However, the portion of the earth that could be "seen" from the burst point would be blanketed by an intense pulse of electromagnetic radiation [EMP]. EMP from high altitude bursts can upset or destroy communications, power and electronic systems over large areas of the Earth. [An entire continent would be affected by bursts from above about 100 kilometers.]

In fact, EMP explosions might well be among the opening salvos in a nuclear war. They could compound the confusion and chaos at a time when critical decisions were being made regarding the use of nuclear weapons. International communications links among political and military leaders could be disrupted and the ability to guide and control military operations in the field could be compromised within minutes after the start of the war. Even the anticipation of EMP effects could have a profound effect in pushing a crisis to the point of war, since combatants might be prompted to act more quickly than they otherwise would—perhaps even pre-emptively—in an attempt to make the most effective use of C^3I facilities before they could be disrupted by enemy EMP.

The EMP pulse reaches its maximum strength almost instantaneously and lasts a few millionths of a second. In this very short time, it can induce damaging electric currents in aerial and buried power and telecommunications systems; in complex systems even a small number of failures could render the system effectively useless. A number of military command facilities are "hardened" against blast, but the extent of their survivability against intense EMP remains at issue. Finally, EMP could represent a threat to increasingly-important space-based systems used for communications, navigation, surveillance and defense.

Effects on Radio Signals

Many satellite and ground-based communications, navigation and sensing systems depend on the reliable transmission of radio waves through the atmosphere. Nuclear explosions in space, in addition to generating EMP, can affect the propagation of many radio frequencies and therefore degrade radio, radar, microwave and satellite signals. The explosions would also alter the nature of the earth's upper atmosphere, particularly the belts of electrically-charged particles that influence communications. Ground-based communications systems which use the ionosphere to bounce signals between widely separated receiving stations could be seriously affected. Blackouts lasting several hours are possible, especially at the frequencies used for short wave communications. How long these systems would be disrupted would depend on how seriously the ionospheric belts were affected by the nuclear explosions in space.

It is perhaps not surprising that, in the past, attention has been focused so strongly on the devastating immediate and local effects of nuclear explosions. A major outcome of this study is to highlight the possibility that a large-scale nuclear war would also have serious long-term and widespread effects beyond the immediate devastation. It finds that non-combatant nations and the global biosphere could not escape grave consequences of such a war. The next chapters will examine these implications, particularly the role that smoke from nuclear-ignited urban fires could play in altering the earth's climate and the impact this might have on global ecosystems, on world agriculture, and on human populations.

ELECTROMAGNETIC PULSE EFFECT

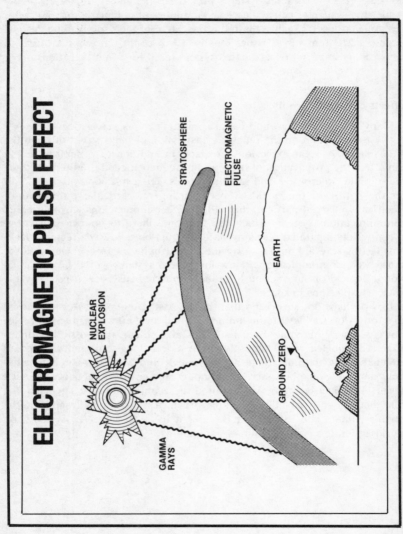

(Adapted from F. Glasstone and P. Dolan, *The Effects of Nuclear Weapons*, U.S. Dept. of Defense and U.S. Energy Research and Development Administration 1977; and from SCOPE 28, *Environmental Consequences*, Vol. I.)

CHAPTER 4
Smoke and Dust

It has long been known that particles in the atmosphere can have important climatic effects, but it was not until the early 1980s that scientists began to suggest that smoke produced by the burning of cities in the aftermath of a nuclear war might significantly affect the Earth's climate for long periods of time after the conflict. Determining the exact nature and extent of these climatic effects has, however, been an extremely complex problem, requiring estimates of:

- the amount of smoke and dust that could be produced by nuclear explosions and nuclear-ignited fires in cities and wildlands. This requires an analysis of the amount and types of materials that would burn and the way in which fires might develop and spread;

- the heights at which the particles would be injected into the atmosphere, which would depend on the intensity of fires and meteorological conditions;

- the nature, number and size of the smoke and dust particles produced;

- the percentage of the particles that would be removed from the atmosphere within hours to days, before the smoke is spread far beyond the initial areas of injection;

- how far the particles could spread around the globe, and therefore how widespread the climatic effects would be;

- how severe and long-lasting the climatic effects might be.

These factors will be examined in detail in the next two chapters.

PARTICLES IN THE ATMOSPHERE

Why the concern about particles produced by burning cities in the aftermath of a nuclear war? After all, cities have occasionally burned to the ground without plunging the Earth into climatic crisis. The answer, in part, is a matter of scale; the large numbers of fires that could be ignited by a massive exchange of nuclear weapons, and their extent and intensity, would be un-precedented in human history. But, even more significantly, fires in modern cities, particularly the burning of large amounts of fossil fuels [e.g., oil and gas] and their derivatives [e.g., plastics, rubber, asphalt] could generate very large amounts of black, sooty smoke. Lofted into the upper atmosphere, this black smoke could have a serious climatic effect by blocking the sun's energy and causing severe cooling at the Earth's surface.

Depending on their amount, physical size, chemical nature and location in the atmosphere, particles can significantly affect transmission of solar energy to the Earth's surface and the temperature distribution and radiant heat flow in the atmosphere. All particles, including the dust that would be kicked up by nuclear explosions, have the effect of *scattering* sunlight, i.e., causing photons [light particles] to be bounced off in varying directions, although mostly downward toward the Earth's surface. However, smoke, which consists of particles produced in the flames of burning organic mate-rials [i.e., materials containing carbon, such as wood and fossil fuels], also *absorbs* solar energy besides scattering it, i.e. solar radiation is transformed into heat. Soot, a component of smoke produced by flaming combustion, is a particularly effective absorber. It is a mixture of oily material and elemen-tal carbon, a very black, almost pure form of carbon that is responsible for the absorption. The blacker the smoke, the higher its content of elemental carbon and thus the more strongly it absorbs solar energy.

As the sun's radiation travels through the atmosphere, it is scattered and absorbed by particles. As a result, the amount of solar energy that reaches the Earth's surface can be greatly reduced. The factor *optical depth* is used to describe this effect; it is a measure of the effectiveness with which the solar energy is intercepted. An optical depth of 1 due only to absorption allows only about 37% of the sun's energy to reach the surface, while ab-sorption optical depths of 2 and 3 respectively allow only about 14% and 5% through.[1] The absorption of this energy causes warming in the atmosphere where the particles are located and cooling at the surface of the Earth, due to a reduction in the amount of energy reaching the surface. Scattering also removes solar energy from the direct rays of the sun, but is not as efficient in reducing the flow of solar radiation to the Earth's surface.

Several factors are important in estimating the effect of smoke particles on the climate: the nature of the fuel burned, fire behavior, the availability of oxygen, the number, size and shape of particles produced, how long they remain in the atmosphere and the height they reach in the atmosphere. The

---

[1]   These figures apply if the sun is directly overhead, in which case the solar radiation takes the shortest path to the surface. More typically, however, the sun will be at an angle. In this case, the solar radiation will travel a longer way through the atmosphere and will be subjected to increased interception by particles, so the amount of sunlight reaching the surface will be even further reduced.

nature of the fuel burned determines how much smoke will be produced and the amount of elemental carbon it will contain. For example, burning wood generally produces less smoke and elemental carbon than burning fossil fuels and their derivatives.

Particle size and shape are significant because they affect the length of time the particles will remain in the atmosphere before being removed. Particles are usually measured in terms of micrometers, i.e., millionths of a meter. Particles between 0.1 and 3 micrometers are the most effective in absorbing and scattering solar radiation; they can remain for long periods in the atmosphere—anywhere from a few days to several weeks in the troposphere to one or two years in the stratosphere. Smaller and larger particles are generally removed from the atmosphere by various processes within a few days. [These removal processes are discussed below.]

The height at which particles are injected into the atmosphere is also very important. The higher the particles, the longer they remain in the atmosphere and the more widely they will be dispersed around the Earth; as a result, the climatic effects would be more widepread and would last longer. The height and dispersion of particles are particularly important factors in estimating the climatic impact in the Southern Hemisphere. Since the Southern Hemisphere presumably would not receive many, if any, nuclear strikes, much less smoke and dust would be produced there, and thus the climatic consequences it would suffer would be largely a by-product of the war in the north. [These geographic factors will be examined more fully in Chapter 5.]

THE PRODUCTION OF SMOKE AND SOOT

Estimating the amount of smoke that could be produced in a nuclear war is a challenging task. Such an estimate depends, firstly, on assumptions made about the nature of the war—in particular, the extent to which urban/industrial complexes, and especially fossil fuel storage facilities, would be hit. As was discussed in Chapter 2, these factors are based on unpredictable human decisions which make it impossible to forecast the exact nature of a nuclear war. Secondly, more information is needed on the composition and amount of smoke—and particularly the fraction of soot—produced by the burning of different materials. Existing data from test burning of small samples in the laboratory may not be applicable to a mass fire situation, although it constitutes most of the information available. Finally, it is necessary to estimate the total amount of materials available for burning in industrial/urban areas and in forest and grasslands; what fraction of these materials might actually burn; and how fires would spread. These factors can vary significantly according to location and season and are therefore also affected by assumptions made about the targeting and timing of explosions.

Urban and Industrial Fires

About 70% of the population of countries in the North Atlantic Treaty Organization [NATO] and the Warsaw Treaty Organization [WTO] live in urban centres. There are about 1000 cities with more than 100,000 inhabitants each. It is assumed in the present study that about 15% of the total available wood and wood-based materials in industrial/urban centers, amounting to some 2000 million tonne, would burn in flaming combustion.

It is also assumed that about 700 million tonne of fossil fuels [oil, gasoline, kerosene etc.] and their derivatives [e.g., plastics, rubber, asphalt, roofing materials etc.] would burn. The amount that would burn in a nuclear war would depend, of course, on the nature and number of targets hit, but, since fossil fuel processing and storage facilities are likely to be high-priority targets, this figure is quite reasonable. This quantity represents about a quarter of all available materials of this kind—including oil, coal, natural gas, asphalt, rubber, plastics and other synthetic materials and industrial chemicals—contained in the cities of the developed countries. These materials are particularly significant because they would produce highly-absorbing sooty smoke and would therefore have the most serious climatic consequences.

Even if only the 120 largest industrial and commercial centers of the combatant nations were affected by the military targeting, this would represent an attack on at least 40% of all the combustible material in the NATO and WTO countries. A high percentage of the combustible materials in cities is concentrated in the central districts, where nearly half the urban population is also located, so the extent to which these central areas would be affected during a nuclear war would have an important bearing on how much material would burn.

The next step is to estimate the total amount of smoke that would be produced from burning materials. These estimates must take into account the fact that different materials produce different fractions of smoke; for example, oil, gas and plastics, when burned, emit more and much sootier smoke containing a higher fraction of elemental carbon than do wood products. The smoke yields for these materials have been estimated from simple test fires [although these estimates may not be entirely applicable to very large fires.] Typically, burning materials convert several percentage of their mass into smoke particles, less for wood and more for oil.

For the range of war scenarios discussed in Chapter 2, it is estimated that the total amount of smoke that would be put into the upper atmosphere from flaming combustion could be between about 50 and 150 million tonne, of which about 30 million tonne would be elemental carbon. As an indication of the significance of this, if these particles were spread over half the Northern Hemisphere, the effective optical depth from absorption by these particles would be about 4 to 5. This means that only about 1% of the sun's energy travelling vertically through this mass of smoke would reach the Earth's surface below. The meteorological, climatic and biological implications will be examined in later chapters.

Forest and Wildland Fires

Cities would not be the only sources of smoke in the aftermath of a nuclear war. Forests, agricultural lands, brush and grasslands would also be incinerated because many high-priority military targets are located in or near such lands.

There are many uncertainties in estimating the extent of forest, brushland and grassland fires. Earlier preliminary studies concluded that a forest fire area of 250,000 square kilometers might be plausible. The season during which the explosions take place is very important; Northern Hemisphere forests and grasslands are much more flammable and susceptible to fire spread between April and October. Effective fire fighting would not be possible in the aftermath of a nuclear war. Assuming that about 20% of the combustible materials in forests would burn, earlier studies concluded that about 30 to 60 million tonne of smoke could be produced from forest fires, of which about 10%—about three to six million tonne—would be elemental carbon. This is much less than the estimated 30 million tonne of elemental carbon that would be put into the atmosphere by fires in cities.

In fact, recent research indicates that previous studies may have overestimated the impact of wild fires. One study, which analysed the actual sites of potential military targets, suggests that the forest area which might burn in summer as a result of nuclear strikes against such targets might be only about a third of that assumed in the earlier studies, although the newer studies also have been challenged by a more thorough land use study. While further research is needed, it would appear that the immediate climatic effects of nuclear-ignited fires in wildlands may be of secondary importance compared with those in industrial/urban areas. However, surface explosions in wildlands would be the source of large quantities of soil dust whose effects would need to be estimated.

Here, it should be noted that, if precipitation were greatly reduced over the longer term after a nuclear war, forests would be subject to increased risk of fire. Moreover, large areas of vegetation killed by climatic and other stresses following the war could substantially increase the fuel available for burning in forests.

THE PRODUCTION OF DUST

The amount of dust created by a nuclear explosion of a particular size depends on how close to the surface the burst occurs and on the properties of the soil located beneath. Atmospheric nuclear weapons tests of the 1950s and 1960s showed that a one-megaton ground burst would raise, on average, between 100,000 and 300,000 tonne of soil. For air bursts, the amount of dust raised is much smaller than for surface bursts and decreases steadily with the altitude of the explosion or *height of burst*. Subsurface explosions displace a greater amount of soil than ground bursts, but most of the soil immediately settles and the dust cloud generally does not rise as much.

Dust is formed in various ways by a nuclear explosion. The high temperatures and pressures of the fireball vaporize and melt quantities of soil and rock, which later solidify into fine, glassy particles. Large amounts of pulverized soil are also blown into the air. The intense thermal radiation of the fireball produces a steam explosion ["popcorning"] of surface soil, and the blast winds churn up dust over a large area adjacent to the burst. The suction and afterwinds created by the rising fireball draw this dust up the stem of the mushroom cloud high into the atmosphere.

In a modern nuclear war, many military sites would likely be hit with more than one weapon. Although there are no data from weapons tests on the effects of multiple bursts close together in space and time, at least one analyst argues that, in an intense multiburst barrage, the amount of dust raised per megaton of yield could be up to 10 times greater than that raised by a single burst. Overlapping fireballs might also carry the dust to higher altitudes than any individual fireball.

The height to which the dust cloud would rise depends on the yield of the weapon and the height of the burst, the local atmospheric conditions and the season. A one-megaton surface or low-altitude burst would loft dust into the lower stratosphere. Because the stratosphere is very stable and small particles can remain there for long periods of time, the amount of fine dust that would reach the stratosphere is important in assessing the long-term climate effects of dust lofted by nuclear explosions.

A computer simulation of a counterforce nuclear exchange involving 2500 megatons of land surface bursts indicates that, 5 days after the explosions, there could be about 40 million tonne of fine dust in the upper atmosphere— roughly the quantity that might be expected from the nuclear exchange scenario outlined in Chapter 2. This calculation suggests that such a dust pall would spread to most of the mid-latitude zones of the Northern Hemisphere. Initially, in some isolated locations, optical depths of 8 or more would occur. Although dust scatters a larger proportion of solar energy, and therefore is not as effective as smoke in blocking the sun's energy from reaching the Earth's surface, optical depths this large would be significant.

This suggests that, even though dust is a much weaker absorber of solar energy than smoke, the quantities of dust that could be raised by a full-scale nuclear war might be large enough to cause climatic effects themselves.

For example, major volcanic eruptions which put similar quantities of fine scattering particles into the stratosphere are often associated with cool global climatic periods and severe weather anomalies.

It should be noted, as well, that dust particles raised by nuclear explosions might be more strongly absorbing than has been previously assumed on the basis of early bomb tests. For obvious reasons, the tests were conducted on coarse, barren light-colored soils in remote places. However, many military targets are located in areas with highly organic soils which, in some cases, are darker. Soil organic material caught in the fireball would be burned, but material raised by the afterwinds generated in the explosion might be lifted into the air in the form of absorbing particles. Typical continental soils are also generally richer in fine particles than soils at the nuclear test sites; therefore, a nuclear explosion in a fine organic soil could produce greater amounts of light-blocking dust than an explosion in a coarse barren soil. The potential contribution of this dust to the absorption of solar energy requires further research. Because almost all of the regions subject to nuclear attack contain vegetation, smoke would be mixed with any dust raised. The combination of dust, smoke and soil organic material would be more potent in causing climatic change than dust alone.

SMOKE AND DUST IN THE UPPER ATMOSPHERE

Rising plumes of hot air from the fires ignited after a nuclear explosion would lift smoke particles high into the atmosphere—but how much smoke, and how high? Answering these questions is a challenging task. We have had little experience with smoke plumes from fires of the intensity and extent of those likely to result from a modern nuclear war. It is difficult to draw analogies from natural wildfires. Forest and grassland fires are fundamentally different, in that they generally start from a single or a small number of ignition points and burn on a narrow front, whereas a nuclear explosion could simultaneously ignite tens or hundreds of square kilometers of forest, creating much larger fires than any observed before.

Even previous experiences in war are of limited value. Some information can be obtained from the firestorms that resulted from the conventional bombings of Tokyo, Dresden and Hamburg during World War II, and from the nuclear bombing of Hiroshima and Nagasaki, but since documented observations are minimal, most of the information is anecdotal. In any event, these fires were small compared to the ones that would be ignited in a large-scale nuclear war; for example, whereas the Hamburg firestorm covered an area of about 12 square kilometers, one of today's large nuclear weapons could simultaneously ignite a city of several hundred square kilometers with greater sources of sooty, smoke-producing fuel.

In any general nuclear exchange, a number of fires would occur in relatively dry environments; in these cases, extensive cloud formation and precipitation would not occur. However, for much of the year in many Northern Hemisphere locations likely to be targeted in a nuclear exchange, atmospheric conditions would be such that smoke rising from large surface fires would enter a cumulonimbus cloud [a towering rain cloud with a flattened "anvil" at the top] whose formation would be triggered and supported by the great amount of heat generated by the fires. Clouds such as this engulfed the plume from the Hiroshima bomb and promptly produced a "black rain" containing soot, dust and ash from fires.

The smoke plumes generated by intense nuclear-ignited fires would be characterized by very strong updrafts and possibly, for fires lasting several hours, strong inward-flowing winds that could fan the fires. It is therefore expected that most of the smoke carried by these updrafts would reach the upper troposphere and the lower stratosphere, unless processes going on inside the cloud can remove these particles. [See the following section for a discussion of removal processes.]

Computer models have been used to predict how plumes would develop from nuclear-ignited surface fires, particularly the heights they would reach and where they would deposit most of the particles. The results of these and other analyses suggest that the effect of rising plumes in depositing smoke particles in the upper atmosphere would be highly sensitive to the local meteorological conditions and the intensity of the fires that result. They also indicate that the pattern of smoke injection would vary substantially according to the season. Because of different seasonal conditions of the atmosphere over probable target areas, more smoke particles are likely to be lofted into the upper troposphere/lower stratosphere in the summer than any other season, with progressively smaller but still important amounts lofted to those altitudes in spring, fall and winter. The computer calculations suggested that about half of the smoke produced by intense nuclear-ignited fires could be lofted to the tropopause in summer.

However, it is important to note that these computer simulations have not yet included all the possible removal processes for smoke that could be going on inside the clouds during the first few days after the nuclear-ignited fires start. The present study, and previous ones, assume that 30% to 50% of the smoke would be removed from the fire-generated cloud systems within a few hours or within the first day.

The removal of this fraction of smoke involves not only the process of prompt removal in thunderstorms. It is possible that, over a range of about 100 to 1000 kilometers, cloud anvils [the flattened upper portion of the cloud] in the vicinity of several closely spaced large fires in highly populated and industrial regions might merge. These anvils might be filled both with soot particles and with large amounts of ice particles formed from water lifted into the clouds. Competing warming and cooling processes in the anvil might lead to further formation of precipitation and smoke removal; however, this outcome is still speculative.

Removal of Particles from the Atmosphere

The extent to which smoke and dust particles are initially removed from the atmosphere is a critical factor in determining climatic impact. Previous studies made simplified assumptions regarding particle removal because there is no alternative way of treating this problem adequately. Recent research efforts have focused on trying to characterize this factor more precisely.

Atmospheric removal of particles—known as *scavenging*—can occur by a variety of means. Fine dust and soot particles, for example, might *agglomerate* or *coagulate* [cluster together by attaching to each other in the air or inside water droplets] to form larger particles that could more readily be removed from the atmosphere by precipitation or by settling. The extent to which this would happen in clouds created by nuclear explosions and resulting fires is uncertain, but it has so far been treated only roughly in some of the models. Long-term coagulation and removal of soot and dust is also a factor that must be accounted for in order to estimate the ultimate climatic impact of these particles. An upper limit to the removal rate of smoke and dust particles can be estimated by assuming that they are removed with perfect efficiency by rain. This has been done in a recent 3-dimensional model study.

Precipitation would be the dominant mechanism for prompt removal of particles from the fire plumes and clouds. Smoke and dust inside clouds could actually affect the rain-producing processes. Water vapor can condense on particles to form rain drops; if this happens with relatively few particles, the resulting water droplets might grow large enough to fall, possibly collecting other smoke or dust particles from lower altitudes. The effectiveness of this scavenging would depend on several, rather poorly understood factors, such as the size and composition of the particles and whether or not they attract water.

Determining the actual fraction of small particles that would be removed by precipitation is a very difficult task because of substantial uncertainties in the knowledge of cloud processes. For example, it is possible that conditions in the clouds might result in the formation of a dense mist, rather than rain because of the simultaneous condensation of water vapor on many smoke and dust particles which compete with each other for condensing water vapor. Another major question concerns the efficiency with which small smoke particles of the size that would cause the major climatic effects would be removed by precipitation [as opposed to larger particles, for example]. There is also a question whether particles produced by the burning of oil storage tanks would be large enough to fall back to Earth by gravitation or by efficient capture by falling rain drops. Existing evidence on these points is, at present, very limited.

As the black rains of Hiroshima and Nagasaki attest, some of the dust and smoke particles could be almost immediately removed by rain. But other evidence, such as observations of forest fires and large oil fires, suggests that a substantial amount of smoke and dust could escape precipitation scavenging and survive to be lofted into the upper atmosphere.

The issue of scavenging remains unresolved. Given the current state of knowledge, the assumption in this study that about 30% to 50% of smoke is removed during the first few days appears to be reasonable.

Three factors are important in estimating removal of particles over periods of weeks to months—the height at which they would be injected into the atmosphere; meteorological conditions; and the size, shape and surface properties of the particles themselves. If substantial numbers of particles were large, they would be more efficiently removed by precipitation or gravity and they might coagulate with smaller particles and remove them as well. Direct observations of wildfire plumes indicate that the particles are predominantly small in size.

It is also important to remember that the presence of the particles could significantly change the meteorological conditions in the atmosphere and this could, in turn, affect their own fate. For example, if large amounts of smoke were to cause heating in the upper atmosphere, it is likely that a very stable condition would be created in which mixing from below, cloud formation and precipitation would be reduced, particularly in the middle and upper troposphere. Under these conditions, particles would remain longer in the atmosphere.

There are large uncertainties in modelling particle removal on long time scales in global 3-dimensional models; however, it is conceivable that the efficiency with which precipitation removes particles may be overestimated. The newest 3-D model calculations, which include precipitation removal [using conservative assumptions] provide new insights. One of the most significant results to emerge from these new calculations is that, under conditions in which atmospheric temperatures have been altered by large amounts of absorbing smoke, there would be a lowering of the region where rain would occur and most of the smoke particles could be lofted above these altitudes. This would permit particles to remain in the atmosphere for a much longer time and the climatic effects could be more widespread and prolonged than otherwise.

CHAPTER 5
Climatic Consequences

Once lofted into the upper atmosphere by smoke plumes, smoke particles would start to spread horizontally and begin to have an effect on the climate. The Earth's climatic system is comprised of five components which vary over widely different time scales:

- The *atmosphere*: the shell of gases that surrounds the planet. It is the most variable component of the Earth's climatic system, changing on a time scale of about a month;

- The *hydrosphere*: the water bodies of the Earth, including oceans, lakes, rivers and groundwater. The oceans are tremendous heat reservoirs and climatic variations occur over a longer time scale than in the atmosphere— months to years in the upper layers to centuries in the deep ocean;

- The *cryosphere*: the ice masses and snow cover of the Earth. Snow cover on land is largely seasonal, but large ice masses, such as glaciers, experience major variations only over periods of hundreds to millions of years;

- The *lithosphere*: the land masses of the Earth. These change over tens to hundreds of millions of years and generally do not interact as much as the other components of climate described above;

- The *biosphere*: the Earth's plants and animals, including humans. Biological organisms respond to climate in a wide variety of ways and many also can have a significant effect on the climate.

The Earth's climatic system has a number of properties and involves a variety of processes. Properties include such things as the temperature of the air, water, ice and land; wind and ocean currents; the movement of ice; air moisture and humidity; cloudiness and the water content of clouds; air pressure and density; ocean salinity and a number of other factors. Examples of processes that connect different components of the climatic system include precipitation and evaporation, radiation and the transfer of heat in various ways.

The large amounts of smoke that could be created in the aftermath of a major nuclear war would be injected into this complex, interconnected climatic system, where they could have an effect on a number of important physical processes. They could, in turn, be themselves influenced by these processes. The nature of these interactions, and the ways in which they could result in large-scale climatic changes, are explored in this chapter.

EFFECTS ON ATMOSPHERIC PROCESSES

Solar Radiation

The injection of large amounts of smoke particles into the atmosphere would have one immediate and vitally important climatic effect: it would radically alter the distribution of solar energy absorption within the Earth–atmosphere system. Under normal conditions, about 30% of solar radiation reaching the Earth is immediately reflected back to space and about 25% is absorbed by the atmosphere. The remaining 45% is absorbed at the Earth's surface and is transferred back to the atmosphere by three processes—as *infrared radiation* [wavelengths longer than visible light]; by upward motion of heat from the warm surface to the atmosphere [*sensible heat*]; and by evaporation of water that later condenses [*latent heat*]. This energy is transferred upwards and horizontally through the atmosphere and is eventually radiated both outward to space and back to further warm the Earth's surface [the "greenhouse effect"].

As described in Chapter 4, particles in the atmosphere may either scatter or absorb solar energy. Most naturally occurring particles, such as water droplets in clouds, ice crystals and soil particles, scatter solar energy, mostly downward. Although they scatter solar energy too, sooty smoke particles are also strongly absorbing and it is this difference that is primarily responsible for the potentially large climatic effects of nuclear-ignited urban and wildland fires.

The presence of a particle layer would also affect the amount of solar radiation reflected back to space by the Earth–atmosphere system. Normally, about 30% is immediately reflected, mostly because of scattering by water clouds, but this amount could be reduced by soot particles and enhanced by dust particles, injected by an exchange of nuclear weapons. Because dust particles are drawn up with the nuclear fireballs that rise to high altitudes, it is likely that the dust particles would form a layer above the soot particles, so that the sun's energy would first be scattered by the dust—a small fraction back to space and the rest absorbed by the soot. For large inputs of smoke, the total amount of solar radiation absorbed by the Earth–atmosphere system could be increased, but, paradoxically, the net effect at the Earth's surface would be cooling, because so much solar energy would be absorbed higher in the atmosphere by the soot that the greenhouse effect is reduced.

Infrared Radiation

As mentioned above, much of the solar radiation absorbed at the surface of the Earth is re-radiated into the atmosphere at longer wavelengths as infrared radiation. This has an important effect on the energy balance of the atmosphere, since radiation to space is the mechanism that cools the Earth–atmosphere system. The greenhouse effect—the blocking of infrared radiation by certain gases, especially water vapor and carbon dioxide—reduces cooling at the Earth's surface. The lower down that the Earth–atmosphere system is heated during the day by the absorption of solar radiation, the more efficient the greenhouse heating effect because more water vapor and carbon dioxide are available above the level where solar absorption takes place to trap outgoing heat radiation from the Earth. The absorption of sunlight at high altitudes by soot particles, therefore, causes a reduction in the greenhouse effect at the Earth's surface. Because there would also be less solar heating, ground temperatures would be much reduced.

Smoke particles could trap some infrared radiation emitted from the surface and cause some warming. This has been included in some 1-dimensional models and these studies indicate that the infrared absorption properties of smoke particles would be unlikely to moderate the surface cooling effect very much. Analysis of the nature of the smoke particles indicates that they would much more effectively prevent solar radiation from reaching the Earth's surface than they would prevent infrared radiation from escaping to space. In other words, their *visible optical depth* would be greater than their *infrared optical depth*. Thus their effect in cooling the Earth's surface would dominate any warming they might cause, except in special circumstances, such as at night and during polar winters.

The 3-dimensional models used to estimate the global climatic effects of atmospheric smoke particles do not yet include these infrared effects. Their calculations might therefore predict too much of a temperature drop too quickly in the early period after a nuclear war when the smoke layer is very dark. Including the infrared effects could cause the *amount* of cooling initially to be moderated but, after dispersion of the smoke clouds within a few days, it would have only a minimal effect on the temperature disturbances.

Day–Night Variations and Day Length

In most climate models, the sun's energy input is not made to vary according to the sun's angle or the differences that occur between night and day. Both factors are ignored and the solar energy is averaged over 24 hours. However, this might not be suitable for studying certain atmospheric conditions that might be possible after a nuclear war—specifically, conditions in which the optical depths in the atmosphere reach levels of about 1 to 2.

In this situation, the sun angle and day/night variations are likely to play an important role. In general, climate models which average the sun's angle underestimate the amount of solar radiation that would reach the surface for intermediate optical depths of 1 to 2. Some current climate models may therefore somewhat overestimate the climatic effects of nuclear war for some time periods and locations. However, over these time periods, the models also have uncertainties due to other factors [e.g. uncertainties associated with removal of particles by precipitation.] It should be noted that, in any event, models which do include the day/night cycle still show large cooling effects

Since many current models average day and night, this effectively eliminates day length as a factor. However, it is important, in biological terms, to include this factor, because many plant species are very sensitive to day length. Computer simulations that included this factor indicate that even under a relatively thin absorbing particle layer, the reduction in day length could be important, particularly at higher latitudes, where the sun's energy follows a long slant path through the atmosphere. [See Chapters 8 and 9.]

Processes Near the Surface of the Earth

The energy absorbed at the Earth's surface under normal conditions is nearly balanced by three major processes that return heat to the atmosphere: infrared radiation, sensible heat, and latent heat. [The relative importance of these factors varies with location and season and there are also substantial differences between the land and the oceans.]

Following the injection of large amounts of soot into the atmosphere, sensible and latent heat transfer over land would essentially turned off. If there were no incoming solar radiation, the ground temperature on land would quickly drop below the air temperature and the normal exchange of heat [from warmer land to cooler air] would be reversed, leading to a cooling of the air in the lowest layers of the troposphere. This is the same as a nocturnal inversion [which occurs normally on clear nights or during the long periods during which polar regions are sunless]. This creates a very stable situation in which downward sensible heat transfer is severely damped, so that the heat supplied to the higher layers of the troposphere cannot be moved downward to warm the Earth's surface.

Heat transfer from soil might contribute to delaying surface temperature drops for a few days, but if the surface were covered with snow any heating effect from the soil would be virtually eliminated.

It is expected that the cooling would also strongly inhibit evaporation. Low evaporation rates plus other factors, such as a temperature inversion over the entire troposphere, would likely result in a large reduction in precipitation over land. [Global precipitation effects are examined more fully later in this chapter.]

In summary, the effect on the Earth's land surfaces of severely reducing incoming solar radiation is to alter drastically the normal heat balance of the atmosphere and continental surfaces. It appears unlikely that heat-exchange processes that operate under normal climatic conditions could successfully counteract the cooling caused by a layer of highly absorbing particles in the upper atmosphere.

Over the oceans, the situation would be quite different because the ocean surface cools much more slowly than the land. As cold air masses moved out from the continents over the ocean, they would be heated rapidly from below. This could cause some precipitation, although light rainfall might be more likely than severe storms because of the overlying warmer layer.

Water Cycle

Further studies are required for a better understanding of the effect that the cooling of the lower atmosphere caused by smoke particles would have on the Earth's complex water cycle. There has been much confusion concerning the effect of water vapor lofted into the upper atmosphere along with smoke particles, either generated by fires or redistributed by fire plumes from the surface. Fires would actually transfer a larger amount of water vapor than particles to the atmosphere, but what is more significant is how much of a *change* in atmospheric conditions this would create. The smoke particles would introduce a very large change in the normal level of atmospheric absorption. Since water vapor is already plentiful in the normal atmosphere and the added amount would be only a relatively small fraction of this except in the storms generated near the fires, the condensed water could be important in removing the smoke. The increase in smoke content would, on the other hand, be enormous, both regionally and globally.

Consensus is lacking on the precise nature of these effects on water vapor. It appears likely that, in an atmosphere altered by the presence of massive quantities of absorbing particles, conditions would be such that the entire water cycle would be reduced in intensity and that evaporation and precipitation would be greatly reduced, particularly over land. What rain or snow does fall would be in general quite light and mainly over the oceans and over coastal regions, although there could be a short-lived increase in rainfall at the southern edge of the smoke cloud [about 30°N] until a large fraction of the smoke has risen and moved further south as a result of solar heating and rising of the air.

Atmospheric Transport and Interactive Effects

The movement of particles in the atmosphere, called *transport*, is a critical factor in assessing their impact on the global climate. If their presence did nothing to affect the atmosphere itself, their movements could be predicted on the basis of existing atmospheric motions; but, as we have seen, the soot particles that would be injected into the atmosphere by urban fires could have a strong heating effect on the atmosphere; this, in turn, could affect atmospheric motions and—coming full circle—the further movements of the particles themselves.

It is likely that the top of the particle cloud would heat from absorption of solar energy and, as a result, particles would be lofted even higher. This effect would be directly related to the amount of sunlight available and thus would vary with season and latitude. The high-level heating together with low-level cooling would cause a very strong inversion between the ground and the upper troposphere, which would tend to prevent mixing of air in the troposphere. Thus, it seems likely that at least the upper part of the particle layer would be transported upward and would exist in a stable atmospheric situation, reducing the likelihood of removal from the atmosphere.

These conditions are very like those that exist in the natural stratosphere; in effect, the particles would create a "stratosphere" of their own. For this reason, the height at which the particles are initially injected might not be as crucial as previously thought. This has an important bearing on the *lifetime* or *residence time* of the particles. In stable stratospheric conditions, particles have typical lifetimes of six months to 2 years, as opposed to a few days or weeks in the more turbulent troposphere. Since the lifetime of particles in the atmosphere has a crucial bearing on the severity and duration of their climatic effects, reducing the uncertainties in this area should be a high priority for further research.

Until recently, computer models have treated the particles as though they do not have this *interactive* or "coupled" transport effect, in which the smoke heating patterns are affected by the winds they generate. Since these early simulations did not accurately reflect the real world, major improvements in the models were sought. One of the important advances on which the findings of this study are based is the development of 3-dimensional models featuring fully interactive smoke transport—that is, models in which particles not only affect the temperature patterns and circulation of the atmosphere, but are then moved around in this "new" atmosphere. This, in turn, alters their further effect on the atmosphere.

Another major improvement in the models is that they simulate a post-war situation in which most of the smoke is injected into the atmosphere from specific regions—e.g. North America, Europe and the Soviet Union—rather than uniformly in a band around the Earth. This more closely approximates what would likely happen in the real world, assuming that most nuclear weapons would be exploded in combatant nations in the Northern Hemisphere and therefore that the smoke clouds would initially be most dense over those areas.

Computer calculations that take the interactive smoke effect and regional smoke injection into account indicate that, for a war that occurs in the Northern Hemisphere during spring or summer, the smoke would be lofted to altitudes of about 10 to 20 kilometers. It would subsequently move southward at these altitudes over the tropics and into the Southern Hemisphere within a matter of a few weeks. For soot injections of 30–40 million tonne, which is within the range considered in this study, the average temperature drop estimated for the mid-latitudes of the Northern Hemisphere would be only slightly less severe when this interactive effect is taken into account than when it is not; at specific points, however, the cooling may be even more severe. At the same time, the new calculations indicate that there could be appreciable cooling at lower latitudes of the Northern Hemisphere and climatic effects in the tropics and the mid-latitudes of the Southern Hemisphere. Further computer modelling studies are needed before the extent of these effects, especially on the Southern Hemisphere, can be described with confidence.

Another major result of the new calculations is the indication of a substantial lowering of the tropopause [the boundary between the troposphere and the stratosphere] in the Northern Hemisphere, with most rainfall occurring only in the lowest few kilometers of the atmosphere over the oceans. The significance of this is that most of the smoke would thus be above the region where precipitation could wash it out. Under these circumstances, the smoke particles would remain in the atmosphere for much longer periods, resulting in more prolonged cooling at the Earth's surface and more time for the particles to be transported to the tropics and Southern Hemisphere where they would also have climatic effects.

It is important to note that these results would apply to a nuclear war that occurred in the northern summer. If the war were to occur in winter, the models indicate that initially there would be little southward movement and possibly faster removal of the smoke particles. The climatic impact on the tropics and the Southern Hemisphere would then be dependent on how high the smoke were initially injected into the atmosphere and whether it remained aloft until the spring, when greater heating could move significant amounts southward. This situation has not yet been simulated in the models.

The computer calculations indicate that the anticipated cooling in the mid- and high-latitude regions of the Northern Hemisphere from spring to autumn would be much more severe, in absolute terms, than that anticipated in the tropics, sub-tropics and the Southern Hemisphere. However, this does not necessarily mean that the biological and human impact of the cooling would be less in the latter regions.

If smoke were injected during the northern summer, the anticipated temperature in the mid and high northern latitudes might not be greater than that encountered in those regions during winter. However, under normal circumstances, winter arrives slowly and predictably, giving plants, animals and humans time to prepare. Therefore, the major impact in the northern regions of a severe cooling after a summer nuclear war would derive from its sudden, unexpected and "out-of-season" onset, particularly if it were to occur in the middle of the crop growing season.

Even though the anticipated temperature drops may seem to be less serious at low latitudes, they could nevertheless imply cooling in tropical and sub-tropical regions which could be accompanied by major decreases in rainfall. Plants in these regions are likely to be even more sensitive to reductions in temperature and precipitation than plants in temperate climates during the growing season; this has important implications for the fate of post-war agriculture. [Agricultural effects are discussed in Chapter 9.]

POST-NUCLEAR WAR TEMPERATURE SCENARIOS

The following tables give computer model estimates of temperature decreases that might occur in various regions of the world after a nuclear war. Such scenarios are conditional on the total amounts of smoke [especially soot] injected, the heights to which the smoke particles were initially injected, the season during which the war takes place and many other factors. Both tables assume injection of about 180 million tonne of smoke containing 30 to 40 tonne of soot into the atmosphere in the middle range of altitudes between 0 and 9 kilometers[1] in the season indicated. The numbers in the tables represent degrees Celsius below normal average seasonal temperatures.

Table 5.1 assumes a war occurring in the Northern Hemisphere during summer, when the climatic consequences in that hemisphere would be most severe. Table 5.2 assumes a war occurring in a northern winter, when the effects would be less serious. The seasons are, of course, reversed in the Southern Hemisphere; the extent to which climatic effects would occur there would depend on how much of the smoke that was lofted into the stratosphere in the Northern Hemisphere would be transported to the Southern Hemisphere and the season during which it would arrive there.

These tables should be interpreted with caution. They are not weather forecasts. Because of the many uncertainties that still limit predictive capabilities, these judgements should not be considered firm predictions of the aftermath of a nuclear war, but rather educated estimates. Still, the tables are intended to present a plausible range of consequences, based on what hard evidence is available. They are offered, despite their uncertain nature, because scientists studying biological impacts—and indeed, anyone concerned

---

[1] This corresponds to the "baseline" case adopted in a 1985 study by the U.S. National Research Council, but in terms of the optical effects of soot, it is consistent with the numbers used in this study.

about the future—need the best available information on which to base their analyses of the environmental consequences of nuclear war.

Table 5.1: Northern Hemisphere Summer War

Region	Phase		
	Acute (1st few wks.)	Intermediate (1–6 months)	Chronic[B] (1st few yrs.)
	(degrees Celsius below seasonal average)		
N. Hem. Mid-Latitude cont. interiors	−15 to −35 Under dense smoke[A]	−5 to −30	0 to −10
N. Hem. sea surface[B] [ice free]	0 to −1	−1 to −3 and local anomalies	0 to −4 and local anomalies
Tropical cont. interiors	0 to −15	0 to −15	0 to −5
N. Hem. coastal areas[B]	Very variable 0 to −5 unless offshore wind when −15 to −35	Very variable −1 to −5 unless offshore wind when −5 to −30	Variable 0 to −5
N. Hem. & tropics small islands[B]	0 to −5	0 to −5	0 to −5
S. Hem. Mid-Latitude cont. interiors	Initial 0 to +5 then 0 to −10 in patches	0 to −15	0 to −5
S. Hem. sea surface[B] [ice free]	0	0 to −2	0 to −4
S. Mid-Lat. coastal areas	0	0 to −15 in offshore winds	0 to −5
S. Hem. small islands	0	0 to −5	0 to −5

Table 5.2: Northern Hemisphere Winter War[C]

| Region | Phase | | |
| | Acute (1st few wks.) | Intermediate (1–6 months) | Chronic[B] (1st few yrs.) |
	(degrees Celsius below seasonal average)		
N. Hem. Mid-Latitude cont. interiors	−15 to −20 Under dense smoke[A]	−0 to −15	0 to −5
N. Hem. sea surface[B] |ice free|	0	0 to −2 and local anomalies	0 to −3 and local anomalies
Tropical cont. interiors	0 to −15	0 to −5	0 to −3
N. Hem. coastal areas[B]	Very variable 0 to −5 unless offshore wind when 0 to −20	Very variable 0 to −5 unless offshore wind when 0 to −15	0 to −3
N. Hem. & tropics small islands[B]	0 to −5	−1 to −5	0 to −5
S. Hem. Mid-Latitude cont. interiors	0	0 to −10	0 to −5
S. Hem. sea surface[B] |ice free|	0	0 to −1	0 to −1
S. Mid-Lat. coastal areas	0	0 to −10 in offshore winds	0 to −5
S. Hem. small islands	0	0 to −5	0 to −5

Notes:

[A] Smoke clouds with an absorption optical depth of 2 or more that remain overhead for several days.

[B] Estimates of climate averages; local anomalies may occur in which temperatures exceed these limits, but these are, at present, unpredictable.

[C] The figures in Table 5.2 allow for a considerable range of variation in the rate of smoke removal in the Northern Hemisphere winter atmosphere.

It is possible that, in the acute phase immediately after the war, the smoke clouds might be very patchy, with some areas much denser than others. This would mean that there could be considerable variation in temperatures and precipitation from location to location. However, in the intermediate and chronic stages, after the smoke cover became more evenly distributed, the day-to-day variability in temperatures and rainfall in most locations might actually be less marked than under normal, pre-war conditions. This would result from several factors, including the fact that the smoke veil would be far more uniform than the patchy cloud cover that would occur under normal conditions.

No estimates were made of changes in precipitation patterns because of the even greater uncertainties still attached to such estimates. This subject, along with several other potentially important environmental consequences of nuclear war, are examined more generally in the next section.

BEYOND THE MODELS

Because modelling studies remain to be done on many important aspects of this problem, the possible environmental effects of a large-scale nuclear war on the oceans, on the monsoons, on coastal regions and small islands, on global rainfall patterns, and on the Southern Hemisphere must be discussed on the basis of our general knowledge of climatic behavior rather than on detailed modelling results.

Oceans

The oceans play an extremely important role in influencing the Earth's climate. If ocean conditions were significantly altered after a nuclear war, this could have a profound effect on the global post-war climate.

It is possible that sea-surface temperatures could be changed, not only by the cooling effect of smoke particles, but by changes in atmospheric temperatures, cloud cover, humidity, wind speeds and other factors. However, there is a huge amount of heat stored in the oceans and it would take very dramatic and long-term climatic disturbances to cause a change in global average sea-surface temperatures of more than a few degrees.

This is indeed what the models used so far to estimate post-nuclear war effects indicate, but these models have only a very crude representation of the ocean surface. Atmospheric models have not so far been linked to detailed models used to study the oceans. A more realistic estimate of post-nuclear war sea-surface temperatures awaits such efforts.

It is likely that there could be significant changes in ocean currents and coastal *upwelling* causing small areas of warming or cooling of up to 10°C or so. [Upwelling is a complicated phenomenon resulting in deep water rising to the surface.] Such effects, if they occured, would be highly location-specific. Changes in surface winds in a post-nuclear war environment might also have a major impact on the ocean state and sea-surface temperatures across the Pacific, as occurs naturally with the so called El Niño phenomenon, which caused major weather disturbances during 1982–3. Such changes could result in significant climatic disturbances in Southern Hemisphere regions such as Australia, New Zealand and southern Africa. Further computer modelling studies are needed to estimate these effects.

Monsoons

Monsoons are large-scale atmospheric circulation patterns driven by seasonal differences between land and sea-surface temperatures. They are responsible for the marked wet and dry seasons across most of tropical Africa, southern Asia and northern Australia. There is little doubt that, in an atmosphere altered by a nuclear war in the northern summer, normal rainfall patterns over these monsoon regions would be drastically affected. The monsoon circulation which provides most of the annual rainfall to the African Sahel, the Indian subcontinent, South-east Asia, China and parts of Japan may be "switched off" in a matter of days by the reversal of the normal Northern Hemisphere land-sea temperature differences. However, a full understanding of these effects must await more detailed model simulations.

Coastal and Island Effects

It is possible only to speculate about the post- nuclear war climatic effects that may occur in coastal regions and on oceanic islands [e.g. the British Isles, Taiwan, New Zealand, Japan, Indonesia, the Phillipines]. However, it is probable that the following may occur:

- coastal areas: cold air flowing from deep continental interior to the warmer coasts and oceans might lead to sub-freezing conditions in coastal areas lasting for days or weeks. This could have a significant impact on crops and on fish and shellfish in estuaries.

- small islands: land surface cooling on small islands might be limited to about 5°C below the surrounding ocean temperatures, with perhaps even less cooling depending on wind conditions.

- Islands and their coasts might experience severely reduced precipitation, depending on the overlying warmer layer and on wind patterns [and, in some cases, changes in these patterns in the post-war atmosphere].

Precipitation Changes

As discussed in the previous chapter, precipitation is not yet well predicted in the global climate models, mainly because precipitation processes occur on a scale too small to be dealt with in detail in global models. Therefore, there are large uncertainties in the estimates of potential precipitation changes that might occur after a nuclear war. Studies with models that generate their own precipitation indicate that there could be substantial reductions in precipitation over most of the world in a post-war environment; one model predicted a 25% drop in rain over land and a 20% drop over the oceans about 3 months after the smoke was injected into the atmosphere.

As mentioned above, it is possible that some monsoon rainfall would fail almost completely and that island rainfall could be significantly reduced. It is also thought that new circulation patterns set up between the Northern and Southern Hemispheres could cause decreased rainfall in southern mid-latitudes. These latter changes might only last a few weeks, however, until a significant amount of smoke had moved into the Southern Hemisphere. There might be increased rainfall over the oceans in many near-shore and coastal regions, but this would probably be steady, light precipitation rather than large storms.

If enough smoke remained in the atmosphere to produce an optical depth of 1 or more after it had spread around the globe, average precipitation could be significantly below normal for an appreciable period. The potential reductions in precipitation could have a major impact on agriculture and natural ecosystems. [See Chapters 9 and 10.]

Southern Hemisphere Effects

Under normal circumstances, it takes a long time—a year or so—to exchange air between the Northern and Southern Hemispheres. However, computer model calculations suggest that, if about 150 million tonne of smoke [or 30 million tonne of soot] were injected into the Northern Hemisphere, changes in the atmospheric circulation could result in the movement of large quantities of particles into the Southern Hemisphere in a few weeks. This would be the major cause of climate disturbances in the Southern Hemisphere upper atmosphere after a war; it is generally assumed that relatively few nuclear weapons would be exploded there and thus the amount of smoke directly created would not be enough by itself to cause significant widespread cooling, although there might initially be local cool spots under thick smoke patches.

Estimates have not been made of the ultimate extent to which the Southern Hemisphere might be covered with smoke and dust after a Northern Hemisphere war. Computer models indicate that the smoke injected in the

Northern Hemisphere might well be lofted above the level where it would be rained out and that it might remain in the atmosphere for long periods— months, perhaps years—and thus would have ample opportunity to mix much more evenly between the hemispheres. It seems reasonable to expect that optical depths in the Southern Hemisphere would continue to increase after the first few weeks and that significant land cooling could occur in southern mid-latitudes.

The season in which the war is assumed to occur has a major bearing on Southern Hemisphere effects. If smoke were injected in the Northern Hemisphere in spring or summer, it would arrive at southern latitudes in the southern winter and would have a minimal effect on surface temperatures. Because of the great altitude of the particles, they could still be in the southern atmosphere through the following spring and summer unless a large proportion was somehow removed by chemical or transport processes. They could then have a greater climatic impact in the southern summer.

It is uncertain what the effect on the Southern Hemisphere would be if a major war were fought in the northern autumn or winter. Much would depend on the height at which the particles were initially injected into the northern atmosphere, which would affect how long the particles would remain in the atmosphere and therefore the amount that would eventually find its way to the Southern Hemisphere. Estimates so far indicate that Southern Hemisphere effects from a Northern Hemsiphere winter war might not be significant unless extremely large quantities of smoke were to be injected into the northern atmosphere.

In the absence of direct cooling caused by significant amounts of smoke in the southern atmosphere, the climate of the Southern Hemisphere could be affected by the changes in atmosphere circulation, wind patterns, ocean currents, precipitation and monsoons that might result from the effects of smoke in the northern atmosphere.

Longer-term Effects

Would the Earth's climate and environment be permanently changed by the after-effects of a large-scale nuclear war? Given the present state of scientific knowledge, this question is impossible to answer with any confidence. It is considered probable that there would be serious global climatic and environmental disturbances for at least weeks and months and that the effects might linger for years. But, on the longer scale of climate, "nuclear winter" would in all probability be a relatively transient disturbance; it need not lead to another ice age. While such a transient disturbance could have devastating biological and human consequences, the Earth might well return to its present normal climate within a few years—unless some climatically significant feedback process occurs that is sufficient to alter the energy balance of the entire planet permanently. Some of the possible feedback processes that might lead to longer-term effects include changes in the surface albedo or reflectivity of the Earth over large areas of burnt or otherwise changed vegetation, and fallout of soot onto snow and ice surfaces. Changes in the amount of moisture transpired into the atmosphere by plants could also have a prolonged effect at least on some local climates. These and other possible effects should be the subject of further research. While the possibility of a more permanent climatic change cannot be completely ruled out, the issue is unresolved.

MODEL UNCERTAINTIES

Climate models have been improved in recent years and they have been demonstrated to be powerful tools for analyzing some climate problems. But this does not guarantee their accuracy on other problems, such as the effect of large amounts of smoke produced by nuclear-ignited fires. Some of the major remaining causes of uncertainties in the models are briefly outlined below:

- the need for better understanding of the physical properties of particles [e.g. size, shape, optical characteristics] and the processes that contribute to their removal from the atmosphere;

- the need to treat several important processes that occur on a scale smaller than the spatial resolution of the models, especially precipitation and scavenging processes and land/sea interactions;

- the need to study more thoroughly the absorption and emission of solar and infrared radiation by smoke and dust particles and to include their scattering properties;

- the need to account for atmosphere/ocean interactions;

- the need for better modelling of chemistry in the post-war atmosphere.

Under normal circumstances, carbon particles are not lofted into the stratosphere and their ultimate fate in a changed, post-war stratosphere is a major question—particularly how they interact with other chemicals, such as ozone, and what effect the higher levels of ultraviolet radiation would have on them. Some of these issues are examined in the next chapter.

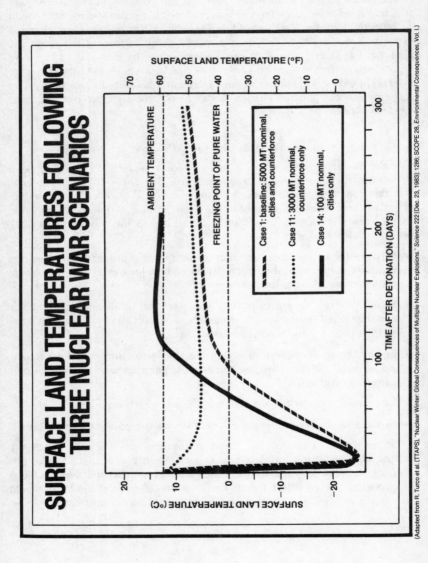

SURFACE LAND TEMPERATURES FOLLOWING THREE NUCLEAR WAR SCENARIOS

AMBIENT TEMPERATURE

FREEZING POINT OF PURE WATER

Case 1: baseline: 5000 MT nominal, cities and counterforce

Case 11: 3000 MT nominal, counterforce only

Case 14: 100 MT nominal, cities only

SURFACE LAND TEMPERATURE (°F)

SURFACE LAND TEMPERATURE (°C)

TIME AFTER DETONATION (DAYS)

(Adapted from R. Turco et al. (TTAPS), "Nuclear Winter: Global Consequences of Multiple Nuclear Explosions." *Science* 222 [Dec. 23, 1983]: 1286; SCOPE 28, *Environmental Consequences*, Vol. I.)

CHAPTER 6

Changes in the Chemistry of the Atmosphere

In the aftermath of a large-scale nuclear war, large amounts of chemical pollutants—including carbon monoxide, hydrocarbons, nitrogen oxides, sulphur oxides, heavy metals, hydrochloric acid and a variety of other toxic chemicals—could be released into the Earth's atmosphere. Some of these substances would be created by the nuclear fireballs themselves and some by the resulting fires; others could be released by the destruction of industrial plants and chemical storage facilities. Some of these pollutants would represent a direct and immediate hazard to plants, animals and humans, especially in regions near their source. Others could have serious longer-term environmental effects; nitrogen oxides, for example, could deplete the ozone layer in the stratosphere that protects life on Earth from the sun's ultraviolet radiation, resulting in an increase of biologically damaging ultraviolet [UV-B] radiation at the surface.

In the densely populated industrialized regions, an important effect might be the build-up of pollutants from smouldering fires following the nuclear exchange, particularly in poorly ventilated cold air masses trapped in river valleys and in lowlands. In these areas, the pollutant levels could greatly exceed levels believed in industrialized countries to be harmful.

It is possible that changes in atmospheric chemistry may occur over months to years after the injection of chemical pollutants. These long-term changes may arise as a result of possible changes to the biosphere [discussed in more detail in the following chapters], although, with the current state of knowledge, these effects remain highly speculative.

CHEMICAL EMISSIONS AND SHORT-LIVED POLLUTANTS

The burning of cities and industrial centers could produce, in addition to the quantities of sooty smoke and dust discussed in earlier chapters, a number of toxic gases and other substances that could cause atmospheric pollution and represent a hazard to human health. These include:

- *Smoke from smouldering combustion*: A few hundred million tonne of "white" smoke, which scatters but does not strongly absorb sunlight, could

53

be put into the atmosphere, remaining at low altitudes. It could contain potentially hazardous organic matter and an appreciable fraction of the particles might be of a size that can be inhaled into the lower respiratory tracts of humans.

- *Carbon monoxide*: It is estimated that between 300 and 900 million tonne of carbon monoxide would be created by fires. The total amount in the atmosphere at present is calculated to be about 1000 million tonne, so the added amount would represent a significant increase over current global levels—perhaps as much as double. Locally or regionally the increases would of course be much larger and could pose a short-term health hazard.

- *Hydrocarbons*: It is estimated that between 40 and 120 million tonne of methane and between 50 and 150 million tonne of other, more reactive, hydrocarbons could be released by nuclear-ignited fires. The amount of methane is negligible compared with levels normally present in the atmosphere, but atmospheric concentrations of the other hydrocarbons would be increased by large factors over existing amounts. The problem could become much worse if hydrocarbons were released from explosions and fires in above-ground fossil fuel storage facilities, natural gas distribution systems and especially from blow-outs of natural gas production wells. This problem was not quantified in this study.

- *Nitrogen oxides*: It is estimated that about 30 million tonne of oxides of nitrogen [NO_x] would be produced by fireballs and urban and industrial fires in a large-scale nuclear war. This is about equal to the amount currently produced annually by automobile exhaust and industrial combustion. The NO_x produced in the nuclear fireballs would have a more important impact on the atmosphere than the industrially-produced NO_x. [See the following section on chemical changes in the atmosphere.]

- *Local concentrations of toxic compounds*: The above-mentioned compounds and many others that might be emitted after a nuclear exchange would not be hazardous if spread over large portions of the atmosphere. It is possible that strong temperature inversions might develop over continental land masses while smouldering combustion continues. If so, some simplified calculations indicate that toxic compounds might be trapped in extended smoke clouds near the ground in densely populated and industrial areas, particularly in river valleys and lowlands at concentrations much higher than legally permitted under normal conditions. Under these circumstances, the density of smoke might become quite high as well, which, combined with possible fog formation, could result in much reduced atmospheric visibilities. The health risk posed by these potentially hazardous events requires further research.

Among the toxic gases emitted, carbon monoxide [CO] would likely be the most serious health hazard, especially near the fires. Although some other gases emitted [e.g., hydrochloric acid and sulphur dioxide] might be considerably more toxic than CO, their emissions would be so much smaller that individually they would not be as significant as CO. However, the simultaneous emission of smoke, CO and other toxic gases can cause human health effects that outweigh their individual effects. Some pollutants, such as sulphur dioxide, are very harmful to plants. [See Chapters 8 and 9.]

Other Emissions and Effects

It is possible that several tens of millions of tons of sulphuric, nitric and hydrochloric acids could be produced from nuclear-ignited fires. If most of these acids were to be rained out within a month over half the Northern Hemisphere, the acidity of rain in general could be increased by about 10 times over the levels currently found in polluted industrial areas. The possible formation of cold acid fogs near the ground could also be especially damaging to biological systems. This requires further investigation.

Chemical factories in areas targeted by nuclear weapons would release chlorine and chlorinated compounds [such as polychlorinated byphenyls or PCBs]. Other toxic substances would be produced from burning these compounds. Spillage and leakage of extremely toxic compounds into soils and water systems could also pose critical environmental problems in heavily populated areas. These problems were too complex to be carefully analyzed in this study. Recent examples of such problems are those that occurred in Seveso, Italy, and Bhopal, India.

Some biological effects of chemical emissions are examined in Chapter 10.

CHEMICAL CHANGES IN THE ATMOSPHERE

What effect would the release of large quantities of toxic gases and other pollutants have on the atmosphere? In the following discussion, a distinction is made between effects in the lower atmosphere [*troposphere*] and the upper atmosphere [*stratosphere*]. In the "normal" atmosphere, conditions in these two regions are very different. The troposphere contains roughly 90% of the mass of the atmosphere. Most of what we know as "weather"—including all rainfall—takes place there. The stratosphere is a much less turbulent place. The atmospheric chemistry that occurs in the two regions is very different and the effect of adding chemical pollutants would vary. It is important to remember again that the chemicals that might be released after a nuclear war would not be going into a "normal" atmosphere, but into a post-war atmosphere in which meteorological conditions might be substantially altered by the presence of large amounts of smoke and dust.

The computer models currently used to study the chemistry of the atmosphere are designed to study atmospheric conditions and chemical processes believed to be important in the normal atmosphere. However, these conditions and processes would be much different in a post-war atmosphere. Existing models cannot adequately handle these changed conditions. It is likely that interactions between various atmospheric chemicals and soot and dust particles could have a major influence on the chemistry of the troposphere. These interactions have barely been studied because they are relatively unimportant in the normal atmosphere.

The Lower Atmosphere

Determining the effect of chemicals released into the troposphere in the aftermath of a large-scale nuclear war is a difficult task. For example, it is not yet possible to tell whether nitrogen oxides produced by fires would lead to a significant increase in the levels of photochemical smog near the surface. Nuclear explosions and fires would produce large amounts of nitric oxide and nitrogen dioxide. When mixed with hydrocarbons in a polluted atmosphere, NO_x can lead to an increase in tropospheric ozone similar to the photochemical smog episodes that frequently occur in industrial areas. However, sunlight is also required for this process, and the amount of solar energy available in the troposphere would likely be reduced by the presence of large amounts of smoke. In fact, unless a substantial fraction of the smoke were washed out almost immediately, chemical reactions on particle surfaces might remove much of the NO_x. Then the formation of excess tropospheric ozone would be unlikely even after the smoke had been removed from the atmosphere.

The Upper Atmosphere

Nitrogen oxides generated by nuclear fireballs and injected into the stratosphere could play a role in depleting the ozone layer which protects life on Earth from the sun's biologically-damaging ultraviolet radiation [UV-B]. A one megaton explosion produces about 5000 tonne of NO. The highest concentrations of ozone, [a form of oxygen consisting of three oxygen atoms] occur at altitudes between about 20 and 35 kilometers. The total amount of ozone in the stratosphere is relatively small but it nevertheless is able to prevent most of the UV-B from reaching the Earth's surface.

Under normal circumstances, ozone is constantly being generated and destroyed by a series of chemical processes. Since the processes of ozone creation and destruction balance each other, the total amount of stratospheric ozone at any one time normally remains about the same. If there were to be a significant increase in the contentrations of chemicals that destroy ozone, the balance between destruction and production could be upset and the ozone layer could be depleted, resulting in an increase in UV-B at the Earth's surface, assuming UV-absorbing smoke had been removed from the atmosphere.

Because of the complexity of the chemical processes in the stratosphere, it is not a simple matter of converting a given increase in stratospheric NO_x into a decrease in ozone; other factors, such as the altitude, transport and dispersion of the NO_x in the stratosphere must be considered. Computer studies indicate that the amount of ozone destruction that could occur after a nuclear war would depend not only on the total megatonnage exploded, but, more importantly, on the yields of the individual weapons used. It is the yield that determines the height to which the NO_x would be raised in the atmosphere. Low-yield weapons do not loft NO_x into the stratosphere at all; even for yields high enough to do so, the amount of ozone depletion depends greatly on the altitudes within the stratosphere at which the NO_x was deposited.

Because there has been a trend toward stockpiling lower-yield weapons and removing the larger yield ones in world nuclear arsenals over the past two decades, newer estimates of ozone reduction are lower than those of a decade ago. In one study, involving a scenario that assumed the use of many multi-megaton weapons from earlier arsenals, a maximum reduction in ozone of 44% occurred after six months, with recovery over several years. For two more recent studies, which assumed the use of smaller-yield weapons projected for current arsenals, the simulations produced maximum globally averaged ozone depletions of 17% and 4%, depending mainly on the assumed mix of weapons yields.

The scenario considered in Chapter 2, which includes some large-yield weapons, would cause ozone depletions that fall between these estimates. They could reach as high as 20% to 30%, with the maximum effect being reached within six months to a year. This could lead to increases in UV-B levels at the Earth's surface of 40% to 100%. An ozone depletion of perhaps several percent could persist for up to a decade. [See Chapter 10 for a discussion of the biological effects of increased UV-B.]

Over more limited areas of the world, especially at mid-latitudes, initial ozone depletion could be much higher. It is even possible to create "ozone holes", characterized by losses of perhaps 70% of ozone over areas the size of the United States lasting up to several days.

Ozone Depletion in a Post-war Atmosphere

It should be emphasized that the above estimates are based on NO_x emissions in an otherwise "normal" atmosphere but do not consider a very different post-war atmosphere in which temperatures and circulation patterns could be altered by the injection of large amounts of smoke. It now seems possible that larger, more widespread, and possibly longer-lasting reductions in stratospheric ozone could result from a large-scale nuclear war. Although the use of lower-yield weapons would mean that NO_x would not be lofted as high in the atmosphere by the fireball, the altered atmospheric circulation patterns could carry not only the NO_x from the fireballs, but also smoke and NO_x from the fires into the altered stratosphere, where both could contribute to ozone destruction. The strong heating of the soot-filled portion of the stratosphere by absorption of solar radiation would also lead to ozone depletion.

It is also likely that smoke created by the nuclear war would change the circulation of the atmosphere so that redistributed ozone in the Northern Hemisphere would promote more ozone destruction. As the smoke spreads to the Southern Hemisphere, the NO_x might spread too, causing ozone destruction there. However, in all these estimates, the role of particles in absorbing UV-B must be taken into account.

Further research is needed to understand more fully the complex interactions among smoke, ozone and other chemicals in the stratosphere and the ways in which they affect and are affected by stratospheric temperatures. Very little information is available on the interaction between soot and dust particles and atmospheric gases. Also, reactions with ozone might provide a method of removing smoke particles from the atmosphere.

In summary, changes in total ozone by several tens of percent would likely occur as a consequence of NO_x emissions into the stratosphere by nuclear explosions, soot particle transport to the stratosphere by altered circulation patterns, and heating of those portions of the stratosphere that contain appreciable quantities of soot. The very presence of smoke and dust in the upper atmosphere might extend the time the ozone layer is depleted beyond the normal 1–2 years. To make better estimates of the amount of ozone depletion that could occur requires the development of more complex 3-D models.

CHAPTER 7
Radiation and Fallout

A large-scale exchange of nuclear weapons would produce unprecedented amounts of radiation that can penetrate and damage biological tissue.

The most intense radiation would be contained in the early or *local* fallout, which would be deposited within the first 24 hours up to hundreds of kilometers downwind from the burst point of a surface or near-surface explosion, depending on weather conditions. This radiation could cause sickness and death in exposed humans and other organisms that might otherwise survive the blast and fire.

Global fallout, which would occur over longer periods, is classed as either intermediate time-scale [occurring within one to 30 days] or long-term [occurring beyond 30 days]. Global fallout would be more widespread but less intense than local fallout; it probably would not cause serious immediate damage in living organisms, but could result in a statistical increase in the incidence of cancers and genetic mutations within human populations over many years.

About half of the radioactivity from a ground burst is deposited on particles large enough to fall out within the first 24 hours; the remaining 50% is on smaller particles that are lofted into the atmosphere and therefore contribute to the global fallout. Ground bursts would be used mainly against hardened targets [e.g. missile silos]. Air bursts are more effective against most other targets. Virtually all of the radiation from an air burst goes into global fallout. Some of this radioactivity may remain airborne for years; since this radiation decays rapidly with time, the longer the radioactive particles remain airborne, the less damaging they will be to biological systems when they fall to the ground.

Lethal radiation doses could thus be delivered in local fallout but are highly unlikely from intermediate and long-term fallout. The *dose* an organism receives is a measure of the total amount of radiation it is exposed to over a given period of time. Unless otherwise indicated, dose levels discussed below refer to gamma radiation delivered external to the body. Additional exposure would be due to beta radiation [which is less penetrating than gamma radiation] and to internal radiation from ingestion and inhalation of radioactive particles in food, water and air [discussed briefly below].

Nuclear health physicists express doses in units called *rads*. A rad can be described as the amount of energy absorbed per unit mass in an organism. An external gamma-ray dose of 450 rads received over a 48-hour period or 600 rads received over two weeks is lethal to about half of a healthy adult population. Few people are expected to survive a dose of 600 rads received in 48 hours. [These figures, it should be emphasized, apply under normal circumstances. The probable shortage of medical facilities and personnel after a nuclear war, the large patient load and the fact that many people would be suffering from other trauma, such as severe injuries and burns, means that more people would be expected to die from the same amount of radiation.]

Lethal dose levels could be produced by local fallout over an area of 1000 square kilometers around a one megaton surface explosion. Dosages from global fallout would be much less and would occur over much larger areas and periods of time. A given amount of radiation is less damaging when it is delivered over longer periods of time.

LOCAL FALLOUT

Estimating the amount of radioactivity that could be created in a large-scale nuclear war, and when and where fallout would occur, is a complex task. As with other environmental consequences of such a war, large variations are possible, depending on the nature of the war, the numbers, yields and height of burst of the weapons used, the location of the explosions, and local conditions at the sites of the explosions. Computer models can be used to estimate, for a given weapon yield, the fallout pattern and the total area subjected to a given dose—for example, a minimum of 450 rads in 48 hours.

Structures and materials can provide a "protection factor" by blocking gamma rays, so people can reduce the external radiation dose they receive by seeking shelter. An undamaged frame house might provide a protection factor of 2 to 3, compared with a factor of 10 to 20 in an intact basement, provided in each case that the people remain protected for a period of 2–14 days. According to computer calculations, the area within which people with a protection factor of 3 would be exposed to at least 450 rads in 48 hours is about one-third of the area within which people with no protection would be exposed to such a radiation dose. Buildings designed with some fallout shielding could provide protection factors of 40 or more, while specially designed fallout shelters could provide factors of up to 10,000.

By a month or more after an explosion, other factors could reduce the effects of fallout on a population, such as weathering [runoff and soil penetration], cleanup efforts, and the ability of the body to repair itself when doses are spread over time or occur at lower rates. Internal exposure and interactions with other stresses are factors that could increase the health effects of radiation.

Not surprisingly, estimating dose patterns and fallout areas is much more complicated when it is assumed that multiple nuclear weapons would be exploded close together, a situation that is likely to occur in a large-scale war. In multi-burst situations, fallout patterns might overlap—that is, their fallout radiation fields could be partially or fully superimposed on each other in ways that are difficult to calculate. Thus, estimates for nuclear war scenarios are sensitive to assumptions made about weapons targeting and local weather conditions.

It is important to remember that, in some regions, there are many potential targets located close together while, in others, the targets may be more widely dispersed. This implies that local fallout could be distributed very unevenly over continental areas, with some regions receiving much higher levels of radiation than others. This would probably be true of the northeastern United States and western Europe.

To provide at least a rough estimate of the effect of multi-weapon bursts, calculations have been made to show the difference between a no-overlap case and a total-overlap case. For one "no-overlap" case, the area of minimum lethal dose was calculated to be twice as large as that in the "total-overlap" case. Other calculations suggest it is likely that, in cases where the overlap in fallout patterns is somewhere *between* the no-overlap and total-overlap situations, the area receiving a given radiation dose could be even greater [by a factor of 3 in one case] than that in the no-overlap case. This is possible since the fallout required to reach a certain level is additive from all the bursts in the vicinity.

A computer calculation of local fallout fields was done, assuming a combination of the no-overlap and total-overlap cases and using the war scenario presented in Chapter 2.[1] The conclusion was that roughly 7% of the land surface of the United States, Europe and the USSR would receive doses exceeding 450 rads in 48 hours. Since there are uncertainties in these calculations, these areas could be somewhat smaller, but they could also be larger. The calculations assume that individuals were not protected against radiation; if higher protection factors were assumed for the entire population, the dose areas would be considerably smaller. Plants and animals and many people would in general have no protection. If a different war scenario that attempted to exacerbate the effects of local fallout were used—for example, surface bursts on cities—the dose areas could be greater and the area affected would be more densely populated. Considering lower dose levels

---

[1] A total of 6000 megatons was assumed to have exploded, of which 1250 megatons would be air bursts and 250 megatons would be ground bursts in urban/industrial areas. The remainder would be evenly divided as air bursts and ground bursts on military targets. See table in Chapter 2.

that are harmful to humans [about 200 rads], even larger areas would be affected.

Also, it should be noted that the distribution of radiation would be affected by wind patterns; for example, the western Soviet Union might receive downwind local radiation from weapons exploded in eastern Europe and the eastern U.S. and Canada might receive local radiation from explosions in the central U.S.

A number of studies carried out by civil defense authorities and defense analysts have predicted that a substantial fraction of the population in the U.S., Europe and the USSR could be under the local fallout plumes. Casualty figures for some scenarios range from less than 10 million to as high as 100 million or more from fallout alone. To the extent that populations could be mobilized to move from highly radioactive zones, or to take substantial protective measures, the human impact of fallout could be reduce. While the present study has not looked specifically into the problem of human casualties, it is not inconsistent with earlier assessments when allowance is made for differing scenarios.

Other factors could increase estimates of the size of local fallout areas. For example, the lower-yield modern weapons may have higher *fission fractions*,[2] which would increase the fallout area per burst. Adding such weapons to the scenario in Chapter 2 could increase fallout areas by roughly 20% above the baseline figure. Also, tactical weapons, which could produce substantial fallout, especially in heavily populated European areas, have been omitted in that scenario. Including these would also increase the total areas. Local fallout might also be produced by explosive dispersal of material in nuclear reactors, nuclear waste storage sites, nuclear processing plants and weapons stockpiles, discussed below. Taking into account all these factors, it is possible that lethal local fallout could blanket large areas and affect large populations within the combatant countries and nearby downwind neighbors.

GLOBAL FALLOUT

Radioactivity can be carried into the Earth's atmosphere on fine particles, which typically remain aloft about a month in the troposphere and much longer if they enter the stratosphere. Particles that remain in the troposphere are typically removed, mainly by rain or snow, within the first month after the explosions and deposited around the globe roughly within the latitude band in which the explosions took place. This is termed *intermediate time-scale global fallout*. The extent to which intermediate time-scale fall-

---

[2] The fission faction is the fraction of the total yield due to fission rather than fusion. All-fission weapons are practical at lower yields. It is the fission process that creates the main radiation hazard.

out would occur after a nuclear war depends on the yields of the weapons exploded, since it is yield that determines how high into the atmosphere an explosion will loft the radioactive particles. The war scenarios used in the most recent studies, including this one, generally assume the use of more low-yield weapons than earlier studies did; this implies a greater potential impact from intermediate global fallout.

Particles lofted into the stratosphere by higher-yield explosions may remain there for years, eventually coming down at points all over the Earth. This is called *long-term global fallout*.

For the same amount of radioactivity lofted into the atmosphere, intermediate fallout from the troposphere can produce doses on the ground about 10 times greater than those produced by long-term global fallout from the stratosphere because radioactive particles remain longer in the stratosphere and have more time to decay before they fall to Earth.

Radiation Doses

To assess global radiation doses that might result from a large-scale war, calculations were done for nuclear war scenarios from two earlier studies; however, assumptions regarding the megatonnage exploded and weapons yields are similar to those used in the scenario cited in Chapter 2. Thus it can be assumed that results for global fallout for that scenario would be similar. The doses are given as the total rads over a 50-year period from external gamma-ray exposure. A majority of the dose would be due to the intermediate time-scale fallout within the first season after the war, followed by a gradual rise to the 50-year level. No account is taken of the effects of weathering of the fallout or of protective measures.

The present estimates indicate that, in an atmosphere not assumed to be changed by the climatic effects of smoke, the entire Northern Hemisphere would receive an average 50-year dose of about 10 to 20 rads; the Southern Hemisphere would receive a dose about 20 times less. In the latitude band between 30°N and 50°N, the average would be about 20 to 60 rads, more than double the overall northern hemispheric average. [In certain land areas of about 1000 kilometers by 1000 kilometers, due to proximity to the explosions and interaction with precipitation and other atmospheric conditions, radiation levels could be on the order of 100 rads.]

Nevertheless, these global levels would not be high enough to cause immediately observable health effects, although they could result in a long-term statistical increase in the incidence of fatal cancers and genetic effects in humans [and possibly radiation effects in ecosystems]. This equates to an increase of several million cancer deaths over the hundreds of millions that would die from cancer without a nuclear war. The same holds for genetic effects.

The estimates of global radiation doses are not significantly affected by the seasonal factor [i.e., whether the war occurred in summer or winter]. The calculated doses are very sensitive to assumptions about weapons yields. For example, when a calculation was done assuming the same total megatonnage exploded, but detonation of a larger number of smaller-yield weapons, the global dose levels were approximately doubled. The calculations done for this study recognize that there are now more low-yield and fewer high-yield weapons than when earlier studies were done about 10 years ago. It is this difference in war scenarios, along with use of more appropriate meteorological and geographical factors, that largely accounts for the current estimates of global radiation dose levels being about 10 times higher than those in the earlier studies.

Post-war Atmosphere

Altered atmospheric circulation and precipitation patterns could have an important effect on fallout patterns and radiation doses. If radioactive particles remain longer in the atmosphere because of these altered circulation patterns, they will be transported farther. One important consequence of this might be an alteration in the relative global radiation doses received in the Northern and Southern Hemispheres.

Calculations that take into account the effect of post-war climatic changes indicate that, in an atmosphere altered by nuclear war, the earlier estimates of overall average global radiation dose in the Northern Hemisphere might be reduced by about 15%. Doses in the Southern Hemisphere would be slightly higher, but the increases would not be large, remaining about 20 times lower than Northern Hemisphere levels. This is mainly because the radioactive particles would take much longer to reach the Southern Hemisphere and be deposited on the ground; therefore the radioactivity would have undergone considerable decay.

Internal Radiation Doses

The radiation doses discussed in the previous section refer only to gamma radiation external to the body. However, in the aftermath of a nuclear war, air, water and food supplies, mianly in the local fallout areas, would likely be heavily contaminated with radioactivity. If consumed or inhaled, the dose would be distributed inside the body.

During the first few months in an area extending hundreds of kilometers downwind of a surface explosion, surface waters might be severely polluted by the debris and radioactivity. Lakes, reservoirs and rivers would gradually become less contaminated as water flows through the system. Ground water supplies might remain unpolluted for years, but they might be difficult to tap. Eventually, however, groundwater could become contaminated in the years after the war and would not return to pre-war conditions for perhaps hundreds or thousands of years.

People exposed to local fallout would have their skin and clothing directly exposed to radioactive dust and, in the post-war chaos, might be unable to find washing facilities to clean themselves off. Some of this dust might also be ingested into the stomach or lungs. Contaminated food and water would also provide internal doses of radiation. Researchers have calculated the doses various organs might receive via many food pathways, but these calculations are questionable for a post-nuclear war scenario involving both changes in diet and the amount of radioactivity consumed.

The resulting internal body dose has been roughly estimated to be about 20% of the external dose for local fallout, about the same as that from intermediate fallout and somewhat greater than that from long-term fallout.

FALLOUT FROM NUCLEAR FACILITIES

The question of whether nuclear power facilities, both military and civilian, would contribute to radioactive fallout after a nuclear war has generated a good deal of controversy. These facilities include nuclear reactors; storage facilities for spent fuel and radioactive wastes; reprocessing plants; and nuclear weapons construction and storage depots. They contain large amounts of radioactive substances; the controversy concerns whether these facilities would be deliberately or inadvertently hit with nuclear weapons and, if they were hit, how widely their radioactive stores could be dispersed into the atmosphere by nuclear explosions.

The present study considered the question of what could happen if such facilities were to be hit—in part because many people have expressed concern about it—and concluded that there could be a three-fold increase in the 50-year global fallout dose estimates if a third of the world's existing civilian nuclear facilities were hit and their radioactivity were dispersed globally. Most of the radioactivity would come from the storage pools rather than the reactor core; it would contribute to the long-term dose and not substantially to the lethal levels of local fallout. If military nuclear power facilities [more likely targets] were hit, they could make a similar contribution to global dose levels.

Nuclear Facilities

Within the nuclear power industry, there are several types of facilities containing radioactive substances:

- *Nuclear reactors*: commercial civilian reactors, military fuel production reactors and reactors aboard naval surface ships and submarines.

- *Storage facilities*: spent reactor fuel rods located in water pools near reactor sites, and other facilities for storage of high-level radioactive wastes.

- *Fuel reprocessing plants*: facilities for recovering radioactive material from spent fuel rods usable as reactor fuel and for the construction of nuclear weapons.

- *Weapons facilities*: for the production, assembly and storage of nuclear warheads.

There is debate whether, if a civilian nuclear reactor were hit, the radioactive core materials could be vaporized and dispersed in the same manner as radioactivity from a nuclear weapon; reactor fuels are highly protected by special coverings, containers and thick walls of concrete and steel. However, if reactors were to be hit with surface-burst weapons, radioactivity in the core could be carried away in the fireball. The reactors attacked could also melt down, if not immediately vaporized, releasing some of their more volatile radioactivity into the local environment.

Spent fuel rods typically are stored in water pools that have comparatively little protection, and their radioactivity would more likely be dispersed if they were located close enough to a reactor that is hit or if they were to be hit themselves. Fuel reprocessing plants contain significant amounts of radioactive materials and are less protected than a reactor core. Finally, reactors on nuclear-powered military ships carry significant amounts of radioactivity and would likely be high-priority targets in a nuclear war.

[It should be noted that nuclear facilities contain a mix of radioactive materials that are very long-lived compared with the mix of radioactivity produced by weapons explosions. This implies that the nuclear facilities would become the major contributor to the additional global radiation doses received after the first year.]

Local Fallout

It is estimated that, in the first 48 hours after a nuclear war, nuclear facilities, if hit, would probably contribute less than 10% of the total external gamma radiation dose in the combatant nations. However, because reactor cores and spent fuel pools contain relatively more long-lived substances than those produced by a nuclear weapon, radioactivity from the nuclear facilities could dominate in the longer term over much larger areas than would be affected by the weapons alone. If the large doses received during the first month, mainly from weapons radioactivity, are substracted out, then, as time increases the deposited radioactivity from nuclear facilities could be the dominant contributor over the longer term, both locally and in areas beyond the local fallout regions.

Global Fallout

The computer calculations used to estimate the potential contribution of nuclear facilities to the 50-year global fallout dose assume an attack on approximately a third of the world's civilian nuclear power industry, or 100 GW(e).[3] The results indicate that the 50-year global doses would be three times greater if such facilities were hit with nuclear weapons than if they were not. Because severe assumptions are made,[4] this should be considered a high estimate; a different set of assumptions could produce either higher or lower estimates. If all civilian nuclear facilities were targeted and vaporized— amounting to about 300 GW(e)—these estimates could go up by a factor of three. Military facilities are considered by some analysts to be equally important and more likely to be hit.

The calculations indicate that the 50-year external gamma-ray dose from a combination of weapons and nuclear facilities would be greatest in the 30°N to 50°N latitude band, with an average value of about 100 rads. This compares with the estimate given previously of about 20–60 when only weapons radioactivity is considered. Levels in the Southern Hemisphere would be about 30 times lower than those in the Northern Hemisphere if nuclear facilities were to be hit.

In summary, there would be a number of potential sources of exposure to radioactivity in the aftermath of a major nuclear war:

- External gamma-ray exposure from local fallout from nuclear detonations;

- Gamma-ray exposure from intermediate time-scale and long-term weapons fallout;

- Beta radiation as a result of close contact with fallout debris;

- Internal exposure from inhalation or ingestion of food and water contaminated with radioactivity;

- Local and global fallout from military and civilian nuclear facilities that might be attacked;

- Long-term contamination of local ecosystems by nuclear reactor debris not lofted into the atmosphere with the nuclear fireballs.

Severe sickness or death could follow early exposure to local fallout. Lower or more extended doses would lead to an increased incidence of cancer and mutations.

---

[3] A GW(e) is a measure of the rate of electrical energy production.

[4] These assumptions are that each facility is hit with a high-yield, accurately-delivered, surface-burst weapon that completely pulverizes or vaporizes all radioactive materials and that these materials follow the same pathways in the environment as radioactivity from the weapons themselves. Destroying the electrical generating capacity of a nuclear power plant would not require these constraining conditions, especially a surface burst.

SURFACE TEMPERATURES: VARIATION FROM NORMAL, AFTER SUMMER NUCLEAR WAR

DAYS 5-10

Surface temperature (°C) variation from normal temperatures
- <-15
- -5 to -15
- > +5

DAYS 35-40

Surface temperature (°C) variation from normal temperatures
- <-15
- -5 to -15
- > +5

Turco, R. P., Toon, O. B., Ackerman, T. P., Pollack, J. B., and Sagan, C. "Nuclear Winter: Three-Dimensional Simulations Including Interactive Transport, Scavenging and Solar Heating of Smoke." (Preprint of paper submitted to

CHAPTER 8
The Biological Response

Mass starvation could well be the most serious human consequence of nuclear war. If the war caused substantial climatic and other environmental disturbances, as discussed in previous chapters, the effects on agriculture and natural ecosystems could ultimately result in the death of hundreds, perhaps thousands, of millions of people who managed to survive the direct effects of the war. Even in the absence of extreme climatic disturbances, disruptions in societal systems, including global trade in food and energy resources, could create severe food crises around the world.

Projections of the climatic consequences that might occur in the first weeks and months after the war, while perhaps the most dramatic, are not a measure of the full human impact. Individuals would not die from a loss of sunlight for a few weeks. Deaths by freezing would not greatly decrease the world's post-war population. It would appear that the Earth's human population has a much greater vulnerability to the effects of those changes on other organisms, particularly when this reduces the availability of food.

In this, as in the climate estimates, there are uncertainties, many of which necessarily result from uncertainties in the climatic projections. There will always be uncertainties about the physical aftermath of a massive nuclear war, for reasons already discussed. To delay assessments of the biological consequences until all the physical uncertainties are resolved means never doing them.

But there is something to work with in making a biological assessment, for it is clearly plausible that serious climatic disturbances could follow nuclear war and, as the following analyses show, one need not assume the worst-case scenarios or the climatic extremes to reach the conclusion that agricultural systems and food supplies around the world could be seriously imperiled by such a war. For example, it is clear that reductions in average temperatures of even a few degrees—the less extreme end of the ranges projected by the climate models—could have devastating effects on agricultural productivity. Thus, it may not be necessary, at least in terms of assessing the impact on humans resulting from food shortages, to reduce further the uncertainties attached to the estimates of the climatic extremes.

Moreover, it is apparent that, even in the absence of any significant climatic disturbances, world agriculture could be seriously disrupted in the aftermath of a nuclear war by the potential loss of international trade; food transportation, storage and distribution facilities; agricultural machinery; energy supplies; fertilizers and pesticides; seed supplies and other subsidies to agriculture provided by humans.

The following chapters examine the *vulnerability* of global biological and human systems to the various climatic disturbances that could follow a nuclear war. In most cases, two time frames are considered: an *acute phase*, the immediate weeks and months after the war, particularly the first post-war growing season; and the *chronic phase*, referring to the longer term, beyond the first year after the war. The relative impact on different biological systems of different climatic stresses—reductions in temperature, rainfall and light levels—are discussed.

Geographic differences are also examined. Different regions of the world would likely experience climatic impacts from the war that differ in kind and intensity and they vary in their vulnerability to the changed climatic conditions.

Because of the uncertainties in projecting the exact nature of a nuclear war and its climatic consequences, the discussion of potential biological effects cannot be taken as a prediction of a certain future. The analysis does, however, suggest that it is within the power of human beings, through nuclear war, to disrupt global ecological and human systems on an unprecedented scale.

This chapter examines the physiological response of plants to climatic changes. In Chapter 9, the vulnerability of world agriculture is considered and Chapter 10 takes a similar approach in examining the vulnerability of natural ecosystems. In Chapter 11, a few aspects of the human impact of nuclear war are discussed—most notably the problem of post-war availability of food.

PLANT RESPONSES TO CLIMATIC CHANGES

We have seen from preceding chapters that a range of significant changes in temperature, precipitation and sunlight levels are estimated to be possible after a major nuclear war. The logical question these projections raise is: how well can the Earth's biological systems cope with these climatic disturbances?

One can start to assess the potential biological impact of the projected changes by examining the response to low temperatures and reduced levels of rainfall and sunlight at the level of individual plants.

Plant Responses to Low Temperatures

Nearly two-thirds of the Earth's land mass experiences mean minimum temperatures below freezing and, for nearly half, the minimum drops below

– 10°C. Clearly, then, freezing temperatures are normal over much of the Earth and many plants and animals have adapted to them. Trees in some northern forests, for example, can survive temperatures of – 50°C. Some northern plant species can even survive being immersed in liquid nitrogen at – 196°C. The major patterns of distribution of plants are determined to a major extent by annual and seasonal low temperatures; however, even within a given ecosystem at a particular location, different organisms can vary significantly in their response to changing climatic conditions.

Many plants can adapt to low temperatures, but these adaptations are geared to fluctuations within a certain normal range that occur regularly and gradually on a seasonal basis. Plants are generally not able to cope with temperature extremes well beyond those normally encountered, particularly if the onset is sudden and out-of-season. In short, plants are prepared for normally-experienced extremes but not for surprises never before encountered. Relatively small changes occuring over many generations could cause the range of plant species to shift geographically, but shorter-term events, occurring over weeks to a few years, can simply cause local and regional extinctions.

Plants respond not only to changes in average seasonal or annual temperatures but to brief episodes of extreme temperatures as well. They can be damaged or killed outright by sudden drops to below freezing at times when they are susceptible and have little opportunity to enhance their cold tolerance. Under normal conditions, many plants *acclimatize or harden*—i.e. gradually prepare to withstand the cold temperatures they normally experience seasonally. Hardened plants are normally not damaged even by freezing episodes; however, extremely severe or exceptionally long, uninterrupted periods of low temperature could represent a danger even to hardened plants and this could be a factor affecting the survival and recovery of plants subjected to an acute nuclear episode.

In contrast to plants capable of surviving temperatures well below freezing, there are many species that are damaged or killed at temperatures well *above* freezing, even up to about 15°C, by what is called *chilling*. A chilling temperature is not cold enough to freeze a plant, but is cool enough to produce injury. [It is thought injury may be caused by direct damage to cell membranes or, more gradually, by upsetting the metabolism of plants.]

Many species in warm temperate, tropical and sub-tropical areas are chill-sensitive. Most do not suffer until the temperature drops below 10°C, but projections of the climatic effects of a nuclear war suggest that temperatures in many regions of the world could drop below this level for days or weeks after a nuclear war. In some cases, temperatures need not drop even this far to produce chilling damage. Some types of rice, for example, suffer damage at 15°C.

The intensity of chill and how long it lasts are important factors; most tropical and sub-tropical plants would be damaged by chilling temperatures within one to a few days. It is estimated that a temperature drop occurring after maize seeds were sown and lasting at least a month would devastate the crop, even if freezing temperatures did not occur.

The resistance of plants to freezing temperatures depends to a large extent on their water content because freezing of water within cells can cause irreversible tissue damage. For example, most micro-organisms, lichens, mosses and ferns can tolerate extremely low temperatures in a dry state. Thus, at the onset of a sudden temperature drop, the survival of many plant species may depend on whether they are dry or wet, or in a growing or dormant state.

It appears that many temperate plants could survive, if in a hardened or dried condition. For example, species common to grasslands and temperate regions, and weedy species, tend to produce dry seed, which can tolerate very low temperatures.

Many tropical and sub-tropical plants are unable to acclimatize to freezing temperatures and are susceptible to cold throughout most of their lives. Many have fleshy seeds and fruits which contain a lot of water and are therefore sensitive to freezing. Plants in warm temperate coastal regions, where frost is rare, also tend to be susceptible.

It should be noted that even if plants can acclimatize to chilling or freezing conditions, this is temporary and reverses if temperatures rise again. For example, in cold temperate regions, plants can "deharden" in 1 to 2 days if temperatures suddenly rise above 10°C. If cold episodes were irregular—resulting, for example, from the passage of dense smoke clouds— acclimatization would be lost during the warmer periods and plants would be as sensitive to damage from subsequent chilling or freezing episodes as they were to the first. A lack of continuous below-freezing conditions could cause susceptibility to cold damage in shrubs and trees, especially in latitudes between 30°N to 40°N.

In general, plants that grow in regions which normally experience seasonal variations in rainfall and temperature have a dormant or inactive state in which they exhibit greater cold tolerance and survivability than plants that grow in regions that are constantly warm and moist. Ecosystems that normally experience little seasonal variation in temperature [e.g. evergreen tropical areas] could be devastated by sudden temperature drops.

Acclimatization and Hardening

Plants can have dramatically different tolerances to low temperatures in their hardened and unhardened states; e.g. in a hardened state, some plants can survive temperatures 20°C to 30°C below those they could survive in their unhardened state.

Plants harden gradually, starting at temperatures around 0°C to 5°C. For many, this is a sequential process; for example, woody species in temperate regions only achieve full hardening after the following sequence: long, warm days; short, warm days; short, cool days; short, frosty days.[1] All steps are essential, and if the short, warm days [autumn] are eliminated, the trees do not harden fully. Trees can be susceptible to sudden freezing temperatures not only in summer but in late spring and early autumn as well, and many woody plants growing north of 40°N to 55°N latitude would be frost-sensitive from April to mid- or late-September.

If a sudden cooling were to occur before September, trees in temperate forests of the Northern Hemisphere could be severely damaged; at a later time, however, most would already be winter-hardened or capable of continuing to harden. It is estimated that if a rapid temperature drop started in mid- September, forests in eastern North America north of 50°N would already be winter-hardened. By mid-October, forests north of 40°N to 45°N would be acclimatized or well on their way and might be able to avoid damage unless temperatures dropped immediately to lows of $-20°$ to $-40°C$. However, the climatic estimates cited in Chapter 5 indicate that temperatures approaching these lows might occur during the acute phase in Northern Hemisphere mid-continental regions.

The timing of cold-hardening differs significantly among different kinds of plants and even among plants of the same species, depending on their location. Northern populations harden earlier than those from southern or coastal regions and, within a species, early winter injury is more common the more southerly the latitude at which they are found. In the Southern Hemisphere, hardy tree species, capable of withstanding freezing temperatures below $-30°C$ have not evolved, perhaps because the winters are milder in the south under the influence of the oceans.

Even plants that are cold-hardened can vary in their susceptibility to extremely low temperatures, depending on the method of hardening they employ. Two mechanisms exist; a hardened plant either *tolerates* or *avoids* ice formation.

Tolerance involves *dehydration* [loss of water] from the plant cells, which would be irreparably damaged if the water inside them froze. Dehydrated plant tissue—e.g. dry lichens, mosses, ferns, seeds and pollen—is extremely hardy. Many species that are regularly subjected to low winter temperatures go into a dehydrated dormant stage and can withstand essentially any freezing conditions without injury. However, this tolerance is a temporary condition, occurring in winter when no plant growth is taking place, and is lost during periods of intensive growth.

---

[1] This does not apply in all cases. For example, the eucalyptus tree and some other woody Southern Hemisphere plants can harden and deharden very rapidly, a mechanism that permits survival during brief, occasional frosts and rapid resumption of growth when temperatures rise again.

In contrast, other plants survive low temperatures by avoiding ice formation through *supercooling* of their tissues [i.e. water in the tissues remains liquid to low temperatures]. Unlike the case with dehydration, the lower limit is about $-40°C$, so plants that employ supercooling are more susceptible to extreme drops in temperature than plants that employ dehydration. Ice avoidance is a mechanism that generally provides protection against relatively mild, episodic frosts in some warm temperate regions.

A severe cooling following a nuclear conflict could result in the loss of plants depending on freezing-avoidance by supercooling. Dormancy and freezing-tolerance brought on by dehydration are the only way plants can cope with severe and long-lasting sub-freezing temperatures.

Most eastern deciduous forest trees in North America have tissues that survive low temperatures by supercooling; if temperatures were reduced below $-40°C$, these trees could be killed or severely damaged. In North America, species that supercool dominate forests in southern Canada and throughout the central and eastern United States. If temperatures were to drop below $-40°C$ for extended periods, or below $-50°C$ for even short periods, these species simply would not be able to survive. Even temperatures of $-20°C$ could kill species below about $34°N$ latitude in the eastern and western U.S.

In contrast, tree species in the boreal forests at more northerly latitudes, where $-40°C$ temperatures occur regularly, survive by ice-tolerance and, if hardened, would be unlikely to die even if the worst post-nuclear war scenario occur.

Germination

The germination of seeds is sensitive to temperature. Seeds will germinate over a wide range of temperatures, from about 3°C to 35°C, but the rate of germination is strongly influenced by temperature. Crops in the cold temperate regions of North America and Eurasia start to germinate at much lower temperatures [1°C to 2°C] than crops grown in warm temperate, tropical or sub-tropical regions, such as rice, which requires at least 10°C for germination.

In ecosystems that are characterized by seasonal changes in temperature and precipitation [e.g., subject to cold and drought], many species have *seed banks*, in which one or more year's worth of seeds are present in viable condition in the soil. Changes in temperature and moisture can often trigger germination of some of these seeds. For many tropical and sub-tropical species that grow in regions with little seasonal variation in temperature, seed banks are rare and they would be much more susceptible to elimination of all seed and seedlings during a catastrophic event than species with large seed banks.

Plant Responses to Changes in Light

It has been estimated that in the acute phase of a climatic change after a nuclear war, sunlight levels could be reduced by as much as 90% and that over the longer term, reductions of 5% to 10% might be possible. However, these changes would be accompanied by temperature decreases, the most sudden and severe occurring during the acute phase, and very low temperatures would have the predominant effect on land plants. For plants that have frozen to death, the lack of sunlight would have little importance. For fresh water plants, the situation would be different; although surface waters might freeze, underneath, the plants would survive. Reduced levels of sunlight would have a direct effect on photosynthesis, so that light levels could become the dominant factor.

Few species are able to survive under continuous low light conditions. However, some plants [e.g. mosses that grow in dense forests] normally survive in deep shade—in some cases, in light intensities less than 1% of that reaching the upper canopy of the trees. It appears that, if a nuclear war caused reduced light intensities for any length of time in the acute phase, shade-tolerant plants could have an advantage over those that occupy open habitats. Some shade-tolerant plants might be able to withstand both freezing temperatures and low light levels in the acute phase, although they might be susceptible to *high* light intensities and lower humidity during the chronic phase.

How long could species tolerate low light, and what sorts of species might prevail under such conditions? There is little evidence on plant survival in the dark, but at least one study indicates that some species are remarkably tolerant of this unnatural condition. Seeds from 25 species taken from open habitats, grasslands and woodlands were studied under darkness with different soil and temperature conditions. The results indicate that woodland species appear to be at a distinct advantage.

Interestingly, plants in soils that were poor in nutrients survived longer than those in nutrient-rich soils, which may have implications for agricultural crops which are normally planted in nutrient-rich soils. In infertile, acidic soil, some seedlings showed a remarkable persistence in dark conditions and were able to resume healthy growth if light conditions improved.

In tropical regions, the temperature drops that might result from a nuclear war are expected to be less severe than in the mid-latitudes of the Northern Hemisphere; in addition, the reduction in sunlight should be less. Many plants in tropical forest are, in any event, adapted to deep shade, but might well be killed by low temperatures during an acute phase. Short-term reductions of a few degrees in temperature and light reductions of a few percent are more likely to be tolerated.

Photosynthesis

Photosynthesis in green plants is the basis of the food chain in all natural and agricultural ecosystems, including those on which humans depend. The maintenance of an ecosystem depends on adequate photosythesis, which, in turn, requires sunlight adequate in both quantity and quality. If light were to be reduced to the levels projected to be possible after a nuclear war, plant photosynthesis would likely be reduced in roughly the same proportion, assuming the plants were actively growing and, perhaps unrealistically, that they were otherwise undamaged. This could have serious consequences throughout the entire food chain.

PLANT STRATEGIES FOR RECOVERY

In the immediate aftermath of a nuclear war, there might be wholesale destruction of established vegetation as a direct consequences of the bombings or subsequent climatic effects. Plants vary in their ability to regenerate under such conditions, but those that maintain a persistent bank of buds or seeds in the soil are likely to be the most successful.

Many plant species that normally go through a life cycle involving periods of destruction and regeneration tend to accumulate large reserves of seeds or spores which persist either in the soil or, more rarely, attached to the parent plants. These can be extremely resistant to temperature fluctuations and may confer considerable post-war resilience on certain plant populations even if there has been large-scale destruction of the established vegetation.

Other species—a large number of trees, shrubs and grasses—form transient seed banks, i.e. they regenerate each year by producing a single crop of seeds which germinate together in the same season after a short period in the soil. This provides a less certain means of survival and the timing of the war would be a crucial factor.

If there is large-scale destruction of the established vegetation, the most vulnerable plants would be those that do not have seed banks, but depend on juveniles [e.g., seedlings or saplings] to regenerate.

Over the longer term, recovery of vegetation would be determined to a large extent by the severity of the initial climatic disturbances and their effect on the survival of different plant species. If the climatic effects were severe, the greatest risk of extinction would occur among relatively stress-tolerant trees in temperate and tropical forests—which may seem paradoxical, since some of these species normally endure large climatic extremes. If the acute effects of low temperatures and reduced light occurred during the winter, hardened trees would indeed be expected to survive. However, a summer war that devastated established vegetation might totally eliminate, at least locally, many species which are unhardened, have no capacity for re-sprouting, and have only transient seed banks or none at all.

In areas where a summer war or close proximity to nuclear detonations resulted in mass destruction of existing vegetation in temperate regions, it is expected that recovery of many species would be slowed by lack of seed banks, poor dispersal ability, slow growth and delayed reproduction. Herbs and shrubs that have several different methods of regeneration would likely expand the fastest, as has been observed in derelict ex-industrial or ex-agricultural areas.

Other unpredictable factors might also come into play. For example, there might be an explosion of insect pests in the absence of bird predators and if application of insecticides on arable land ceases. This could have a considerable impact in the early recovery process, changing the mix and dominance of species in post-war ecosystems.

It is tempting to suggest that the major impact of post-war climatic effects on vegetation would be felt in the acute phase, but this may not be true for some ecosystems. For example, it is expected that many Mediterranean species would survive the acute phase as bulbs or dormant seeds, only to be devastated later in the chronic phase by potential changes in the alternating wet-cool and hot-dry seasons to which they appear to be rather inflexibly adapted.

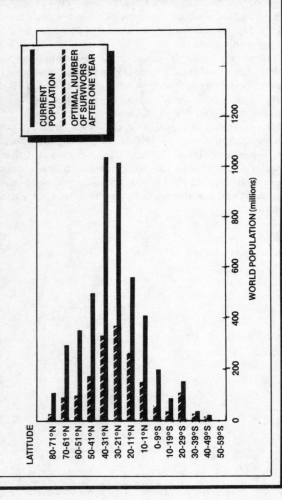

VULNERABILITY OF HUMAN POPULATION TO LOSS OF FOOD PRODUCTION

LATITUDE

CURRENT POPULATION

OPTIMAL NUMBER OF SURVIVORS AFTER ONE YEAR

WORLD POPULATION (millions)

80-71°N
70-61°N
60-51°N
50-41°N
40-31°N
30-21°N
20-11°N
10-1°N
0-9°S
10-19°S
20-29°S
30-39°S
40-49°S
50-59°S

(Adapted from SCOPE 28, *Environmental Consequences*, Vol. II.)

CHAPTER 9
Agriculture after Nuclear War

Without agriculture, less than 1–10% of the nearly 5,000 million people now living on Earth could survive indefinitely. Therefore, the fate of hundreds of millions of people, especially those in non-combatant nations, would likely be determined by the availability of food following a major nuclear war.

It appears, however, that the viability of most of the world's major agricultural regions could be seriously threatened by potential climatic changes triggered by such a war, even those in the less severe range of estimates. It does not require acceptance of a "worst-case" climate scenario to raise serious concern about the fate of world agriculture. It is also possible that climatic disturbances could continue to affect crops in the period beyond the first growing season after the war.

Most agricultural plants, even those widely grown in mid and high latitudes, originally evolved in tropical or sub-tropical environments. As a result, they can be very sensitive to even brief cold episodes, insufficient water and disruptions in their life cycles. A dramatic increase in agricultural productivity has resulted largely from the artificial environment provided by humans—including, for example, protection from predators, pests, disease and competition from other plants; the addition of nutrients and water where needed; selective breeding; and some protection from the natural elements. These support supplies are often referred to as *human subsidies*.

The weather remains the most important uncontrollable variable. Since different crops respond differently to changes in climatic and other conditions, the distribution of a variety of crops over different locations reduces the risk of failure when adverse climatic variations occur. Serious global-scale climatic disturbances are very rare. When climatic factors disrupt agriculture on a local or regional basis, this often can be compensated by trade from other regions although this does not always happen in the case of poorer nations. Thus, the complex world agriculture system, with inputs of human support, has been largely successful in maintaining current populations levels—*but only because the scale of climatic disturbances has been relatively small compared with the global scale of the agricultural system*. The recent Sahelian famine has demonstrated the fragility of *regional* agricultural and food distribution systems.

A major nuclear war would change all this. It might, for the first time in human history, cause global-scale climatic changes that could simultaneously disrupt agricultural productivity world-wide. There is simply no precedent for the the extent and scale of stresses that could be imposed on agriculture by nuclear war. No other human activity has the potential to do this, which largely accounts for the fact that such an occurrence has received little previous scientific investigation. Studies done prior to consideration of potential climatic disturbances would not lead one to believe that agricultural effects would be dominant after a nuclear war; the present analysis, on the other hand, suggests that lack of food could be the most devastating human impact of such a war.

This chapter examines the effect on agriculture resulting from potential climatic disturbances that could be caused by a large-scale nuclear war. The climatic changes considered are consistent with the ranges cited in earlier chapters and considered to be plausible by atmospheric modellers. However, because of the uncertainties associated with the climatic estimates, it must be emphasized that the following discussions of potential biological effects cannot be taken as forecasts or predictions of what will happen. Rather, the approach is to consider the *vulnerability* of various agricultural systems to climatic disturbances of the kind and intensity estimated to be possible.

Throughout the discussion, three major classes of climate disturbances are considered:

- *Acute temperature reductions*: Short-term, extreme reductions in average land or surface-air temperatures to near- or below-freezing levels in the early period immediately after the war and associated reductions in light by 90% to 99% below normal;

- *Chronic temperature reductions*: Longer-term reductions in the average annual surface-air temperatures of a few degrees below normal, and an associated reduction in light by 5% to 20% below normal;

- *Precipitation*: Acute or chronic phase precipitation reductions of 50% or more below annual averages.

THE VULNERABILITY OF CROPS TO CLIMATIC CHANGES

On a global scale and over long periods of time, the average temperature of the Earth remains extremely stable. For example, over the past century, the average annual temperature of the Northern Hemisphere has varied by *less than 1°C* from one year to the next. The difference between current hemispheric average temperatures and the warmest periods on Earth in the last half-million years is only 2°C to 3°C. Yet, these differences, though *seemingly* small, can be accompanied by serious climatic consequences. The climatic estimates cited in Chapter 5 must be viewed in this context.

Those estimates are generally expressed in terms of averages that apply over long periods of time and large regions—for example, reductions in the average seasonal temperature over the mid-continents of the Northern Hemisphere. To understand what this really means, however, it is important to distinguish between average climatic conditions and actual weather at a given location at a particular time, which is how most people, and plants and animals, experience weather.

Under normal circumstances, it is not unusual for cold waves to last a few days on a local or regional basis; for example, on a given day, the local temperature could be, say, 15°C below normal. But, when climatic conditions are averaged over longer periods of time and/or larger areas, they would not decrease by such large amounts; it would be very unusual, in the case cited above, for the local temperature *averaged over the entire year* to be more than a couple of degrees below normal. The same reasoning applies to location. A particular locality or region might experience an unusually cold month, but the average temperature of an entire continent or hemisphere, under normal circumstances, is not likely to drop by as much.

Thus, the potential drop of, for example, 35°C during the first few post-war weeks in the average seasonal temperatures of the mid-latitude continental regions of the Northern Hemisphere represents a climatic disturbance of unprecedented magnitude. We must not dismiss the seriousness of such a change simply because we may personally experience a seemingly equivalent effect in a single location at a given time without serious consequence.

Estimates of potential changes in post-war precipitation patterns over large regions of the Earth must also be considered in the light of normal variations. In large portions of the agriculturally productive regions of North and South America, Europe and Asia, the annual average rainfall varies less than 20%, although more arid regions can experience larger deviations. It is estimated that average rainfall might be reduced by as much as that 20% to 50%, at least on a regional basis, after a nuclear war.

Thus, unprecedented changes in average annual rainfall could occur in the most agriculturally productive regions on Earth. Most grain crops, which provide the major source of human food, are grown in water-limited areas and would be especially vulnerable to reduced precipitation.

The response of biological systems to climatic changes must generally be evaluated on much smaller scales of time and place than those used for the climate estimates given in Chapter 5. Both the cumulative temperature reduction over a growing season and the occurrence of brief episodes of very cold temperatures would be important.

Estimates of temperature changes that are averaged over large areas or extended periods of time are not reliable indicators of the true biological consequences of such temperature disturbances. Two examples will demonstrate this point, using an average drop of 3°C for illustration: Suppose first that every day, the maximum and minimum temperatures were lower by 3°C. The effect of this would be a loss of *degree-days*[1] over the growing season, with a consequent effect on crop yields, and a reduction in the length of the growing season, defined as the frost-free period, or the number of consecutive days with a minimum temperature above 0°C. It is estimated that, at U.S. latitudes, there could be a 10-day reduction in the frost-free period for every 1°C drop in the average growing season temperature. The reduction in the growing season could result in total crop losses if the plants did not have sufficient time to mature, even if the number of degree-days were adequate.

But there is another, different climatic situation that could result in a 3°C drop in average temperature. Instead of every day being a little colder, it's possible that there could be more frequent, more intense, or longer-lasting cold episodes within the normal growing season. In this case, even though the average temperature drop is the same as in the previous case, the biological consequences might be much more serious. Whereas, in the first case, many plants might be able to withstand the relatively small drop in daily temperatures, in the second case, they might be killed or irreversibly damaged by brief episodes of very cold or freezing temperatures. Some studies indicate that brief, extreme episodes are very important in determining the probability of crop failure; the risk of such failure was markedly greater than the shift in average temperature alone would suggest.

With these considerations in mind, the potential response of different agricultural systems to a climate altered by nuclear war can be discussed in some detail, examining, in each case, both the acute and chronic periods after a nuclear war.

NORTHERN TEMPERATE AGRICULTURE

The mid-latitudes of the Northern Hemisphere are among the most agriculturally-productive regions on Earth; their production includes a very large fraction of all the wheat, maize [corn], and rice grown.

Many of these growing areas are located in countries that are among the most likely to be directly hit in a large-scale nuclear exchange. Moreover, analysis of the potential climatic consequences of such a war suggests that these northern mid-latitude regions would also be the most likely to suffer from acute, extreme conditions immediately after the war.

---

[1] The concept of *degree-days* suggests that the time it takes a plant to develop depends on how far above and how long temperatures remain above a certain critical minimum level or threshold. Above the threshold, the rate of growth is directly proportional to the temperature rise. Below the minimum, growth and development stops.

The Acute Phase

The degree to which Northern Hemisphere agriculture would be affected can be determined by knowing how large an area would experience freezing episodes. The climate estimates presented in Chapter 5 suggest that most inland areas of the northern mid-latitudes could experience sub-freezing conditions. For post-war climatic disturbances that occurred in spring, summer or early autumn, any sub-freezing episodes that did occur would coincide with the growing season and essentially no agricultural crop production would occur that year.

It is concluded, then, that acute, extreme climate disturbances occurring after a nuclear war could cause crop losses for the year in all northern temperate regions in which they occur. For the more extreme scenario estimated in Chapter 5, agricultural productivity would be lost if the war occurred anywhere within a six-month "window of vulnerability" from spring through early fall. Moreover, significant productivity might be unlikely after a large-scale war occurring at any other time of year.

In addition, if sunlight levels in the acute stage were reduced by 90 to 99% of normal, crops as a whole would receive insufficient light levels for photosynthesis. Reductions in precipitation could also be a significant limiting factor in situations where the temperature reductions are less extreme.

The Chronic Phase

The climate estimates indicate that, in the chronic phase, average growing season temperatures could be below normal by several degrees—1°C, 3°C, 5°C or even 10°C, with associated light reductions of, respectively, about 5%, 13%,16% and 26%. On a regional scale, precipitation might be reduced 20% to 50%, although local increases in rainfall are possible.

What are the potential consequences of these stresses on agricultural productivity? There is no historical precedent for the extreme climate disruptions that might occur during the acute phase after a nuclear war. Even some of the climatic disturbances estimated to be possible in the chronic stage are more severe than have ever occurred in the last few centuries. Still, the historical record provides some useful insights.

Consider, for example, 1816—"the year without a summer." In the northeast U.S. it was characterized by abnormally low average summer temperatures, which delayed the onset of summer conditions. More significantly, several brief extreme cold episodes and night-time frosts occurred in June, July and late August. The latter effectively terminated the growing season by two weeks. The impact of these transient episodes on crops was devastating. The situation was made worse by the fact that the reduction in average summer temperatures prolonged the time it took for crops to mature, so that they were in a less consumable state when the early onset of frost occurred. As a result, grain prices increased and, while there were no reported food shortages in North America, areas in Europe suffered famine and social disruption.

A second historical analog is the "Little Ice Age", a period extending for about a century in the 1700s during which there were prolonged periods of regional-scale adverse weather conditions. Scientists have used anecdotal evidence to deduce an estimated 1°C drop in average summer temperatures and a 3–4 week reduction in the average growing season. In parts of Europe, grain yields fell by up to 75% over the period and human population growth was reportedly slowed or even reversed in many areas, apparently the result of increased grain prices which put food beyond the resources of poorer people.

Another approach to understanding the effects of chronic climatic stresses on agriculture is to expose plants to various stresses in the laboratory and assess their physiological responses. A study based on this type of experimentation suggests that a 3°C drop in the average growing season temperature would destroy tropical and vegetable crops and reduce the northern range of temperate crops. Similar studies on the effects of reduced light levels suggest that, if light levels fell below about 10% of normal for prolonged periods, most crop plants would die from an inability to photosynthesize and subsequent exhaustion of their energy reserves.

In the context of nuclear war-induced climatic disturbances, these studies suggest that the estimated reductions in sunlight of 5–25% would not reduce crop yields by more than the equivalent fractions and therefore would seem to be minor compared to the effects of the temperature reduction. However, this conclusion will be reconsidered below.

A third method of evaluating the chronic climate phase is the use of computer simulation models. A study of this sort was done to evaluate the impact of climatic stresses on wheat and barley production in western Canada, one of the major grain-producing regions of the world. The model was used to assess reductions in seasonal temperatures of 1°C, 2°C, 3°C, 4°C, and 5°C. Precipitation was assumed to range from 25% above to 25% below normal. In one of the cases studied, the average temperature reduction was calculated in a way that reduced the length of the growing season by delaying the frost-free period and moving up the onset of the first fall freeze. The simulation showed that a 2°C drop in the average growing season temperature would eliminate most western Canadian wheat production and a 3°C drop would completely eliminate it. Barley would be less sensitive, but would be almost eliminated in Canada by a 4°C drop.

These reductions stem from two factors. First, the growing season was reduced by 7 to 10 days. Second, the decrease in average temperature reduced the rate of crop development and growth, thereby exacerbating the effect of the reduction in the length of the growing season. The model predicts that the time required for plants to mature increased by 3 to 5 days per °C reduction in temperature.

The simulations also suggest that, if rainfall were reduced by 25% [without temperatures being reduced], barley and wheat yields would be cut by about 40%. The studies further suggest that a 50% reduction in growing season rainfall would essentially eliminate wheat production in Canada.

The computer model was also used to simulate the reponse of crops to the passing of a smoke cloud created in the aftermath of a nuclear war. In this case, the seasonal average values for temperature and sunlight not only were below normal, but continued to decrease over a period of three months. It was assumed that temperatures were reduced by 5°C the first month, by 3 °C the second month and by 1°C the third month, accompanied by reduced light levels of 11%, 8% and 3% respectively.

The studies showed that, if the temperature reductions began in any month between May and August, spring wheat crops in most of western Canadian would be devasted. For temperature reductions beginning in June, virtually all Canadian wheat production would cease. Reductions in light levels of 11% or less would not cause major losses, but higher levels [e.g. 22%] occurring in the middle of the growing season could result in almost total losses.

The Canadian barley crop would be more resistant to both temperature and light reductions, but combinations of these effects could have a greater impact. The occurrence of peak events in June would devastate the crops.

These studies suggest the following conclusions:

- Chronic reductions in average growing season temperatures of 2°C for spring wheat and 4°C for barley would eliminate these crops in western Canada regardless of any changes in precipitation or light levels.

- The growing season would be reduced by 7 to 10 days at the same time as the time required for plants to mature would be increased by 3 to 6 days.

- In the case of transient cooling episodes, crops are very sensitive to the time when the transient episode of cool temperatures begins.

- Reductions in the amount of solar energy and day length would be important above a certain level; a 10% reduction might result in no crop losses but a 20% reduction, depending on timing, might result in total losses.

- The combined effects of temperature and light reductions would exacerbate the impact on crops.

These results provide the most convincing evidence to date that even if the climatic consequences of a nuclear war were small, they could have a tremendous impact on agricultural productivity. The results also indicate for the first time that, if even relatively small climatic effects were to linger from an acute phase immediately after the war into the next spring, this could cause substantial or complete crop losses in the following growing season.

Soybean Crops

Soybeans could be an extremely important crop after a nuclear war because they are self-pollinating, they do not require synthetic fertilizers and they have a high protein content which could be used to replace protein from animal sources.

Simulation studies indicate that, for the most part, reductions in the average growing season temperature [without changes in precipitation] would reduce soybean crop yields in midwestern U.S. locations by 10% to 25% for a 2°C drop and by 30% to 55% for a 4°C drop. A 6°C drop would likely result in total crop loss. It was concluded that, in these regions, soybean production would be impossible during the acute phase after a nuclear war.

These simulations are an important complement to the Canadian grain simulations. Together, they suggest a high vulnerability of both cold-climate and warm-climate crops to a reduction of only a few degrees in average growing season temperatures. This sensitivity seems to be even greater for the world's most important food, rice. [See tropical agriculture.]

Summarizing these and other studies, it can be concluded that:

- a nuclear war causing significant climatic disturbances between February and August in the Northern Hemisphere would probably eliminate any useful crop yields in northern temperate zones;

- climatic disturbances after crop harvesting could linger into the spring and potentially limit crop yields from the next growing season as well;

- a 5°C reduction in average growing season temperatures would result in extremely low yields in essentially all the wheat-growing regions of the Northern Hemisphere. A 2°C reduction could reduce the area in which wheat could mature by 50%;

- a 50% reduction in precipitation would cause substantial wheat crop losses in the U.S. and the USSR because these crops are already rainfall limited;

- maize productivity would be limited by the lack of hybrid seeds after a war. Temperature reductions of 2°C to 5°C would decrease yields in virtually all areas and eliminate them in northern regions. Southern regions could, however, have high crop yields.

- temperature reductions between 2°C to 5°C could greatly reduce or eliminate crops in many potato-growing regions of Europe and the USSR.

• these are assessments of the *biological potential* for post-war crop growth and production. Other factors, including societal and economic disruptions, could combine to ensure that even this reduced post-war potential could not be achieved and that total production of major crops might well be only a fraction of current levels, even assuming the least severe climatic consequences.

TROPICAL AGRICULTURE

A disproportionate number of those who survive the immediate effects of a large-scale Northern Hemisphere nuclear war would likely live in tropical and sub-tropical regions. Thus, the ability of tropical agriculture to withstand the potential climatic effects of the war would have an important bearing on longer-term human survival.

Plants in lowland tropical regions rarely experience freezing conditions, so direct evidence on the effects of such an occurrence is limited. Laboratory studies and records of infrequent freezing events in locations such as Florida provide some insights. Five Florida freezes between 1977 and 1985 caused huge losses of vegetable crops and fruit. Damaged trees typically took several years to recover.

Most tropical and sub-tropical agricultural plants are sensitive to chilling injury and are killed by brief freezing episodes. During certain stages of its life-cycle, rice can be damaged by temperatures as high as 15°C. Available evidence suggests that brief climatic events producing minimum daily temperatures of 10°C to 15°C would substantially reduce yields of most crops grown in tropical regions. At lower temperature, production could be eliminated. Reductions in the average growing season temperature beween 3°C to 10°C would also cause widepread yield decreases and could reduce the total land area suitable for certain crops. For example, it has been calculated that a 3°C drop in growing season temperatures in Brazil could reduce the area suitable for growing soybeans, coffee and sugar by 165 million square kilometers.

Reductions in precipitation could be a problem in the tropics. Arid and semi-arid regions would be especially vulnerable but even areas that normally receive a lot of rain could be affected if the onset and duration of rainy seasons were to change.

Rice

More than 400 million tonne of rice are produced annually. This grain currently supplies more than 1000 million people with more than half their energy requirements and an important portion of their protein intake. After a nuclear war, a disproportionate fraction of the survivors in non-combatant countries would probably live in regions that currently produce rice; thus, if it could be grown, rice would likely be the major cereal crop available to those survivors.

Rice is grown in China, Japan, Southeast Asia, Indonesia, the U.S., the USSR, Eygpt and to a much lesser extent in South America and Africa. About 45% of the world production of rice is grown under irrigation and another 40% consist of rainfed crops. [Deepwater and floating rice constitute the remainder.] The Asian production is the largest and most significant in terms of human food supplies.

Precipitation is critical to rice production. Most of tropical southeast Asia receives enough rain per year for at least a single crop but the two largest rice-growing nations in the world—China and India—receive less rain. Almost all rice in China is grown under irrigation. In India, crops are often adversely affected by either too much or too little rainfall. Levels of sunlight are also important to rice crops, though not nearly as critical as precipitation.

In more northerly latitudes, rice crops are affected by low temperatures, which can slow or stunt growth, cause sterility or abnormal development and generally interfere with the plant's life cycle. Japanese historical data suggest that a 1°C to 2°C drop in average growing season temperatures could result in crop failure; single-night episodes below 15°C could destroy at least a third of the crop.

It is clear that, if sub-freezing temperatures occurred during the growing season, rice crops would be killed outright and even with temperatures well above freezing, the current growing season crop could be lost. Thus, in the acute phase, rice production could be eliminated in the Northern Hemisphere while Southern Hemisphere crops might experience a similar fate, depending on the temperature changes that took place there.

If growing season temperatures during the chronic phase dropped below 15°C, total crop losses could be expected to continue. Various simulation and other studies indicate that a chronic reduction of 2°C to 3°C in average growing season temperatures would eliminate Northern Hemisphere rice production. In the Southern Hemisphere, the post-war temperature effects might be less severe, but the rice crops would be more sensitive to changes in temperature and precipitation; thus, it would appear that these crops would also be vulnerable to nuclear war-induced climatic disturbances.

In areas with sophisticated irrigation systems, rice production might survive reduced precipitation conditions, if the timing of the war were such that water stores were already built up; otherwise crop production could be affected by reductions in precipitation. Irrigation systems could not be sustained if rainfall were reduced by 50% over a longer period. In regions without sophisticated irrigation systems or where certain types of rainfed rice are grown, reductions in precipitation could cause substantial reductions in yields.

Prolonged periods of reduced light levels could cause reduced rice yields, but this is likely to be secondary to the effects of reduced temperature and precipitation.

SOUTHERN HEMISPHERE MID-LATITUDE AGRICULTURE

As discussed in Chapter 5, if virtually all weapons used in a large-scale nuclear war were exploded in the Northern Hemisphere, the potential climatic consequences would be different in the Southern Hemisphere than in the Northern Hemisphere. Southern Hemisphere effects would likely be less severe, although still potentially significant for agricultural production.

Agriculture in many mid-latitude regions of the Southern Hemisphere [including Australia, New Zealand, southern Africa and southern South America] differs from that in northern mid-latitude because the dominant agricultural activity is pastoral farming for wool and meat production.

Although southern mid-latitudes are not expected to experience a postwar acute phase as severe as that in the Northern Hemisphere, frosts might occur in the growing season that could damage horticultural crops. However, the more dominant grasslands and cereal crops would be largely able to withstand such an event.

Potential chronic phase effects in Australia have been studied with a computer model, assuming 2–5°C temperature reductions accompanied by a 50% reduction in precipitation. The results indicate that the precipitation effects would be the most serious, with biomass reductions of 25%. The estimated temperature reductions were significant only in the regions of Australia that are normally the coldest.

Wheat production in Australia could also be most affected by changes in rainfall. Some studies show that Australian crops might be very sensitive to climatic effects caused by alterations in atmospheric/oceanic circulation patterns.

From these studies, it can be concluded that precipitation reductions would have the dominant effect on crops in Australia [and, by extension, southern Africa and northern Argentina]. It is estimated that Australian grassland and cereal crop production could decrease between 0 and 50% at the beginning of the chronic phase. If stresses last only a year, production could return to normal but might subsequently drop because of a lack of fertilizers. Nevertheless, Australia could still be a self-sufficient food producer, but with considerably lower production.

In contrast, New Zealand's crops likely would respond more to potential temperature decreases, the timing of the effects being very significant. The maximum impact would occur in the spring or early summer. Grassland yields could be reduced, which in turn would cause livestock feeding problems. It is thought that animal numbers could decrease by 10% to 30% and that overall food production could decline by about 20% to 50%. Since New Zealand agriculture currently supports more than twice the population, it would be able to sustain its present population through several years after the war even with the reductions estimated.

It is important to remember that the various environmental effects of nuclear war would not act on crops independently, but rather interactively. In some cases, one stress could cause a reduction in the sensitivity of crops to another stress. For example, under some circumstances, lower temperatures might cause some crops to need less water, making them less vulnerable to a reduction in precipitation. However, it is far more typical for the combination of effects to magnify their impact on crops—i.e. to be *synergistic*. The potential for synergism strongly suggests that the food problems after a nuclear war would be worse, rather than better, than the calculations based on responses to individual stresses might indicate. These synergisms could include not only environmental and climatic effects but societal ones as well. They have not been adequately studied, but some possible examples include:

- crops could become more vulnerable to disease and pests when subject to stresses from radiation and air pollution;

- temperature decreases could reduce populations of insects that pollinate crop plants;

- overexploitation of crops and the environment could reduce long-term agricultural productivity [e.g. harvesting consumable parts of plants before the crop matures];

- inadequate diets may increase the incidence and susceptibility of humans to disease;

- a loss of weather and crop information could coincide with increased uncertainty about future climatic conditions;

- social disruptions could affect the availability of human labor for agriculture;

- social disruptions could interfere with the optimum distribution of food that would ensure survival of the maximum number of people.

It would appear that global agricultural productivity is more vulnerable to the potential after-effects of a major nuclear war than has generally been believed. This conclusion does not require acceptance of worst-case predictions about the environmental effects of such a war. Analysis has shown that even relatively small climatic disturbances could disrupt agricultural productivity on a large scale. For that matter, disruptions of social and economic systems, even in the absence of climatic effects, could be adequate to produce the same results. These issues are further examined in Chapter 11.

In Chapter 10, the response of natural ecosystems to potential nuclear war-induced climatic changes is explored. It seems probable that, in the face of likely post-war disruptions in agricultural and social systems, survivors might increasingly turn to these natural ecological systems for food, fuel, shelter and other means of support but these systems could not sustain human populations at anywhere near current levels.

CHAPTER 10
Ecosystems after Nuclear War

The Earth is divided into a number of *biomes*—a term used to denote a biological community dominant over broad regions of the Earth. In most cases, the term specifies not only a plant community, but its geographic placement—e.g. "north temperate moist forest" or "tropical desert scrubland". Biomes have typical temperature and precipitation regimes and plant and animal species have adapted to these conditions.

The greatest proportion of the Earth's land surface [16%] is covered by biologically unproductive desert, rock, sand and ice. Tropical rainforests cover about 11% of the land, but account for nearly a third of all the primary biological production[1] on Earth, a third of animal biomass and nearly half of the plant biomass.

Water covers 71% of the Earth, but marine ecosystems account for only about a third of primary production and less than 1% of the world's plant biomass; however, they do contain about half the animal biomass. Among marine biomes, the open ocean predominates but it is less productive than other regions such as the continental shelf areas, estuaries, or reef systems.

It is estimated that about 30% to 40% of the Earth's land surface would be within the latitudes most likely to be directly affected by nuclear detonations. The biomes located in these regions contribute an estimated 30% of the total world net primary production, 30% to 40% of the total world plant biomass and about 25% of the total animal biomass.

The temperature and precipitation disturbances, described in Chapter 5, that are possible in the aftermath of a nuclear war could disrupt natural systems on the biome scale. These potential changes are examined, taking an approach similar to that adopted in the analysis of agricultural systems in Chapter 9—i.e., an attempt is made to evaluate the *vulnerability* of various ecosystems to the range of possible climatic stresses.

Consideration is given in each case to acute and chronic phases after both a summer and winter war. To summarize briefly, a summer war would have a much greater impact on ecosystems than a winter war, assuming the climatic disturbances estimated in Chapter 5. These could be expected to be much greater in summer than in winter. Plants and animals adapted to normal winter extremes would not be in a cold-hardened state.

---

[1] Primary producers are organisms, mainly plant, that can produce organic material directly by using the sun's energy for photosynthesis. This organic material provides the energy [food] base for all the higher levels of the food chain [i.e., herbivores and carnivores].

The evaluation is based partly on the physiological responses of plants to reduced temperature, precipitation and light, as discussed in Chapter 8, and partly on modelling studies. The model simulations are considered to be reliable guidelines for assessing the vulnerability of ecosystems to the relatively milder stresses that might occur during the chronic phase, but the results should not be taken as precise predictions.

NORTHERN TEMPERATE ECOSYSTEMS

Arctic and Boreal Ecosystems

Acute post-war climatic disturbances occurring in autumn or winter would have minimal effect on northern forests and arctic ecosystems, which would be winter-hardened and dormant, able to withstand temperatures down to $-50°C$ to $-70°C$, well below those projected for a nuclear winter. If winter conditions were extended into the following spring and summer, the effect on dormant plants would probably be small, unless the environment warmed sufficiently to release the plants from their cold-hardened state. An extended winter could, however, result in major losses of arctic wildlife.

If the acute climatic effects of a nuclear war were to occur during the summer, the impact could be much more significant. Many trees, shrubs, grasses and other plants would be actively growing and therefore potentially vulnerable to subfreezing temperatures. In some regions, a high mortality rate could be expected. Little is known about the cold-hardiness of arctic plants in summer, but it is possible that nuclear winter conditions might trigger rapid cold-hardening. Many arctic species maintain large seed banks and would likely survive.

Populations of arctic animals could be reduced if the plants on which they depend were killed, damaged or buried in snow. Many herbivores, mammals and birds would be unable to survive until the following summer for lack of food. Some species could, in fact, be completely eliminated.

In the chronic phase, plants in the sub-Arctic could experience a prolonged period of winter dormancy and the start of the following growing season could therefore be delayed by up to a month. Plants would probably start to grow, but might not be able to complete a normal life cycle. The extent and frequency of fires might increase if large numbers of trees were killed during the acute phase. Animal populations would be severely affected by lack of food. In the high Arctic, persistent snow cover associated with chronic temperature reductions could effectively eliminate growth during the summer, but plants that remained cold-hardened would likely survive over the long-term. Land animals and birds, however, could starve within a few months.

Forest Ecosystems

The forests of the Northern Hemisphere temperate regions are located at latitudes in which a large-scale nuclear war is considered most likely to occur, so they could experience the most intense stresses—fire, cold, drought, radiation and locally high concentrations of toxic gases. Again, a summer nuclear war would have more severe effects than a winter nuclear war.

In winter, northern trees would be dormant and thus able to withstand acute climate conditions. In more southerly locations, trees, especially pines, would be less cold-hardy and perhaps one- to three-quarters of the trees, shrubs and herbs could be killed, at least above ground. Reduced light levels would not be expected to have a major impact because of low or non-existent photosynthesis during this period. The relatively high density of seeds in forest ecosystems would promote recovery.

If sub-freezing temperatures persisted for two or more months after a summer war, most trees, shrubs and herbs would die. Mammal and bird populations could be eliminated by temperature extremes and lack of food. In following years, forests with substantial amounts of dead matter above ground could be subject to increased insect outbreaks and fires, extending stresses into the chronic phase even in the absence of continuing climatic effects.

In the chronic phase, plants could be killed or damaged if the growing season were shortened or sub-freezing temperatures occurred during the growing season. Reduced light levels could reduce forest productivity. If precipitation were reduced, the danger of fires would be substantially increased.

Simulation modelling of forest systems has shown that a drop of 3°C in average temperature over a 5-year chronic phase resulted in a loss of 25% of the forest biomass, with recovery over 3 to 4 decades. With a 6°C drop in average temperature, 80% of the forest biomass was lost, and 50 years after temperatures returned to post-war levels, the forest biomass was still less than half of what it would have been had the 5-year chronic climate stress not occurred. Other simulations indicated that a 6°C temperature drop or more could change the mix of tree species in the forest and that the forest would be less sensitive to even a 25% reduction in precipitation than to the 6°C temperature drop.

These simulations suggest that chronic-phase disturbances alone could affect forest productivity for decades after the climate returned to normal. Damage that might occur during the acute phase or as a result of increased fires could retard recovery considerably. It could be expected that these changes in forest ecosystems would translate into even greater impacts on animal populations.

Temperate Grasslands

The major grasslands of the Northern Hemisphere are located largely in mid-continental, mid-latitude regions likely to experience most severe climatic and other environmental consequences of a nuclear war. In the acute stage, a winter nuclear war would have the greatest impact on large mammals and birds rather than on plants. The acute effects of a summer war could be expected to have a severe impact on grasslands and their animal populations throughout North America, the USSR and Asia.

In the chronic phase, even relatively small temperature reductions could have a serious impact on production and food availability for animals. Model simulations suggest that a chronic-phase temperature reduction of several degrees could shorten the growing season, which could, in turn, significantly affect populations of grazing animals.

ARID AND SEMI-ARID ECOSYSTEMS

Arid and semi-arid regions extend from the Arctic to the sub-tropics and they can be cool or warm. Warm semi-arid regions have a greater diversity of plant and animal life than either cool semi-arid regions or deserts, which occur primarily in sub-tropical regions.

In the acute phase, a winter war in cool semi-arid regions would not be expected to have a great impact on plants which would normally experience cold temperatures in winter. Hot deserts, on the other hand, could be significantly affected, since many plants would not be cold hardened. Large animal, bird and small mammal populations would also likely suffer.

If a summer war caused sub-freezing temperatures and low light, significant acute-phase impacts on plants and animals could be expected in both cool and warm semi-arid regions.

In the chronic phase, precipitation effects would be dominant. In some cases, the reduction in temperature could result in an increase in primary production, but if precipitation were reduced 25% to 50% at the same time, this would cause a net decrease in productivity. However, major species in the ecosystem probably would survive because they are adapted to long periods of drought. Species in semi-arid regions should be less sensitive to episodes of extreme cold temperatures because they often experience large temperature drops between day and night and thus are better adapted than, for example, forests.

TROPICAL ECOSYSTEMS

The severe post nuclear-war climatic effects estimated to be possible in the Northern Hemisphere are unprecedented in the tropics; therefore, there is little in the way of natural climatic parallels to guide our assessment of the potential tropical impact of such a war. There are also uncertainties concerning how severe, extensive, rapid and long-lived climatic effects in the tropics would be.

The vulnerability of a wide range of tropical ecosystems is estimated for an acute phase characterized by freezing temperatures lasting for days or weeks and a chronic phase, with temperature reduction between 1°C to 5°C. Other factors considered include: reductions in precipitation from 25% to 100%; susceptibility of tropical ecosystems to drought-induced fires; reductions in light levels of up to 90% in the acute phase and 1% to 10% in the chronic phase; possible responses to increased coastal storms and hurricanes; and, finally, differences in response between wet and dry seasons where they occur.

In general, because temperatures are relatively constant in tropical ecosystems at low altitudes, episodes of extreme cold can be devastating in those regions. Because low temperatures can occur in high altitude tropical ecosystems, the impact of a sudden cooling may not be so severe in them.

Tropical Rainforest [Evergreen Forest]

The rainforest is a uniform environment in temperature, precipitation and light levels and is rich in plant and animal species. Their plants are very sensitive to chilling temperature and it is estimated that acute-phase freezing or sub-freezing episodes would kill all vegetation above ground. Recovery could occur if the freezing episode were limited to no more than a week. Plant species [e.g. bamboo] that sprout from below or near the ground might be best able to survive. Many trees and shrubs would be vulnerable because they have limited seed banks; seedlings and saplings would be killed by cold along with the parent trees. After such an acute phase, it is likely that the rich diversity of species would be reduced.

An acute phase with freezing temperatures is also likely to take a considerable toll on animal populations, many of which have poor resistence to cold and might suffer from lack of food. Monkeys, larger animals, fruit-eating birds and many others would not be expected to survive. Hibernating animals would not have an advantage because the onset of cold temperatures would likely be too sudden. Among the species that might survive are carnivores like vultures and eagles; some insects; and perhaps reptiles and shore and migratory birds.

Deciduous [Seasonal] Forests

These forests would be most vulnerable to post-war climatic disturbances during the wet season, when effects could be as serious as those for evergreen forests. During the dry season many species are dormant and the effects of freezing would not be as great, although extinctions could occur. If the freezing were prolonged or occurred repeatedly during the growing season, all existing vegetation would die. Recovery would be more rapid in this type of forest because of better seed survival, larger seed banks and better sprouting ability.

As with the evergreen forest, animals species would suffer greatly, especially from a freeze in the wet season. Many bird and insect species could be severely affected, including migratory birds whose summer habitats could be devastated by climatic disturbances.

Reduced light levels would be secondary to the temperature effects.

If extreme cold during the acute phase killed much of the vegetation, and precipitation were reduced, deciduous forest could be subject to fires much more severe than those which occur naturally. This could destroy the seed bank, retarding recovery.

Tropical Alpine Ecosystems

Wet tropical alpine ecosystems are almost perpetually wet, misty and cold; plants growing there are adapted to a day-night cycle of freezing and thawing. These systems might not be completely devastated in the acute phase if the cold did not last more than a couple of days and did not reach the lower extremes of temperature estimated to be possible.

Tropical Grasslands and Savannah

Grasslands and savannah are widely distributed through South America, Australia, Africa and Indo-Malaysia. Grasses would be most vulnerable during the wet growing season, when a few days of sub-freezing temperature could kill above-ground vegetation. During the dry season, grasses would be more protected and damage would be limited.

Large mammals, especially grazing animals, could die from cold and lack of food in the acute phase. Carnivores would likely have more food and could survive if the cold does not last for more than a few weeks. Reptiles, amphibians and many bird and insect species could also die, although mobile species might migrate.

On the whole, tropical grassland-savannah regions are likely to be more able to withstand potential post-war climate stresses better than evergreen tropical forests or mangroves.

Mangroves

Mangroves are swampy areas in tropical and sub-tropical coastal regions. They are important feeding and nursery grounds for shellfish [e.g. shrimp] and fish species. They are also important feeding grounds for birds and provide a home for many species of fiddler crabs and mammals such as the manatee.

Mangroves would be among the most sensitive ecosystems to the kinds of climatic disturbances projected to occur after a nuclear war; they would be very vulnerable to temperature drops, to reduced precipitation and to increased storm activity along coasts. Evidence indicates that recovery from such storms would be slow; in Florida, for example, damage from a storm in the Everglades in 1934 is still visible after more than 50 years.

Mangrove plants do not adapt to cold, do not have seed banks and could be almost completely killed in the acute phase by sub-freezing or chilling temperatures. There could be extensive deaths among fish, shellfish and plankton.

Chronic Phase Effects on Tropical Ecosystems

In the chronic phase, if temperatures persisted at about 10°C below normal, the productivity of most tropical plant species would decrease by more than 90%, but some recovery of systems would be possible from seed banks if precipitation levels were not reduced below 25%. However, it is expected that the mix of both animals and plant species might be altered. If reduced precipitation were prolonged, the risk of very damaging fires would increase.

In general, tropical ecosystems would be less affected by reduced light levels than reduced temperatures.

FRESHWATER ECOSYSTEMS

Freshwater ecosystems include ponds, lakes, streams and rivers, and range in size from a farm pond to the Great Lakes, a small river stream to the Amazon. It is expected that a post-war reduction in temperature and precipitation would cause a reduction in water levels in rivers and lakes. Most of the Earth's surface fresh water is stored in lakes, so the following discussion focuses primarily on them. Although temperature effects would be less pronounced for any large bodies of water compared with land areas, lake ecosystems would be more vulnerable to post-war temperature effects than ocean ecosystems. One of the most serious potential consequences of a nuclear war for both humans and animals is the freezing of surface waters.

If air temperatures stayed below freezing for a long period of time, a thick ice layer would form on the surface of a lake. The thicker the ice, the longer it remains, restricting human access and possibly creating a low-oxygen environment that could be lethal to some aquatic organisms. Thick ice on a shallow lake can also tie up a significant fraction of its volume.

Under normal circumstances, northern lakes can develop a thick ice cover [up to 2 meters] that can last for about six months between fall and spring. Most biological activity takes place during the early part of the open-water period, starting in about May. There is also a rapid build-up of animal populations in the spring and summer.

Fish follow a definite pattern. After the low-food period during winter, they must restore body energy. During the summer, they feed, grow and reproduce; in the fall, they store body fat as energy reserves for the following winter. Disruptions in this cycle could lead to the loss of juveniles and perhaps even adults. Here again, the impact of losing a year-class of juveniles would depend on how long-lived the adults are and how often they reproduce.

Biological effects would follow if temperate lakes froze over in spring-summer or if more southerly lakes froze over at any time. Some calculations suggest that Northern Hemisphere lakes could develop an ice cover of 0.5–1.2 meters as a result of post-war climatic disturbances, depending on the war scenario. These calculations did not account for potential differences related to seasons, latitudes or continental-vs-coastal effects, so the actual extent of freezing requires further study. Statistics suggest that the largest number of lakes—and therefore those most available to humans—belong to the smallest and shallowest categories, which would have the largest fraction of their water volume tied up in ice if they were to freeze.

Again, model simulations indicate that biological effects would be more severe following a summer war than a winter one—phytoplankton and some of the species that depend on them for food could be lost for the year—but indications are that the lake ecosystems would recover rapidly in either case.

If the nuclear war were to occur in winter in regions where lakes normally freeze over, the ice cover could be thicker than normal; coupled with low light levels, this might reduce photosynthesis and oxygen levels in the lake to the point of killing the phytoplankton. In shallow lakes, freezing could reach the bottom, killing most biota in the lake. In lakes that normally do not freeze over, a freezing episode could totally disrupt biological activity.

The impact of chronic effects would depend a great deal on the time of onset. If spring melting of ice were delayed by lingering climatic disturbances, many animals would not have time to complete their life cycles. [2] In the case of a late spring war, or freezing in southern lakes at any time, there could be wholesale elimination of all organisms from direct effects of low temperature and light. Summer freezing might not be immediately as serious, since many species would have passed their most sensitive life stages, but the outlook for survivors would not be good, since they likely would not have enough food to get through the following winter. In northern lakes, onset of climatic disturbances in the fall would have the least consequences because most organisms would have begun acclimatizing to winter conditions. For more southerly lakes, however, the effects could still be severe.

---

[2]   In this connection, when the spring melt finally occurred, it could be accompanied by a very large acid pulse, as acidic particles drained into the lakes.

SOUTHERN HEMISPHERE ECOSYSTEMS

It is expected that climatic disturbances in the Southern Hemisphere would be less severe than those in the Northern Hemisphere. In large part this is because it is assumed that most nuclear weapons' explosions would take place in the Northern Hemisphere. Another important factor is that climatic effects in the Southern Hemisphere are more subject to the great moderating influence of the oceans, which cover a much larger proportion of the Southern Hemisphere than the Northern Hemisphere. There could also be a time lag before the climatic effects occurred in the Southern Hemisphere because the transport of smoke particles across hemispheres would require massive changes in atmospheric circulation patterns and the movement of particles across thousands of kilometers. Finally, the potential loss of the monsoons could have a devastating effect on the tropics and extratropics.

The potential vulnerability of ecosystems in Australia and New Zealand to post-war climatic changes was assessed. In addition to rangeland/grasslands and oceans, previously discussed, forests, deserts and freshwater ecosystems were also considered. Australia has both warm and cool semi-arid regions and true, extreme deserts. Since most of these ecosystems are routinely subjected to prolonged periods of low precipitation, they could cope with the levels of reduced precipitation projected to be possible in the post-war chronic phase. It is also expected that none of these regions would be vulnerable to brief acute periods of reduced temperature and low light levels, or to longer periods of milder climatic disturbances in the chronic phase.

Effects on freshwater ecosystems could also be limited because temperatures do not seem likely to remain below freezing sufficiently long for these ecosystems to freeze. If there were a prolonged period of low precipitation, long-term water availability might become a problem for humans.

Ecosystems in New Zealand would appear to be vulnerable to sub-freezing temperatures, rainfall reduced to half of normal and other extreme climatic disturbances, but it does not seem plausible that these extremes would occur there after a Northern Hemisphere nuclear war.

Many indigenous species are both drought- and frost-sensitive and it is felt that brief temperature episodes near or below freezing could cause substantial plant death. It is difficult to project the likelihood of freezing temperatures. Although, as a small island, New Zealand would be more influenced by the moderating effects of the ocean, it has a great range of altitudes and more temperate climate. Reductions in precipitation could become a problem in some regions and freshwater streams might be particularly affected. It is difficult to project the post-war precipitation regime, however, because New Zealand's precipitation is controlled by on-shore winds from the ocean, which cannot be adequately modelled at present.

MARINE AND ESTUARINE ECOSYSTEMS

Marine ecosystems include those located in the open oceans, both near the surface and at great depths; on the continental shelves; near the shore [e.g. mud flats and beaches] and in estuaries. A number of these ecosystems support animal species that currently provide major food sources for humans. The fate of these animals in the aftermath of a nuclear war could have some impact on human survivors of the war.

Much less is known about marine ecosystems, but some potential post-war impacts can be suggested. It must be remembered that post-war physical changes in large water masses would not be as large as those projected to be possible on land and they could occur after a considerable time delay. The effects in coastal regions could be highly variable, depending on the effects of the war in adjacent land areas. It is also likely that, in many marine ecosystems, the projected reductions in light levels might have a more important biological impact than the temperature reductions. They latter would be more moderate in most marine environments than they are estimated to be on land.

Open Ocean Ecosystems

The open oceans contain communities of plankton, large swimming animals, crustaceans, and a variety of other organisms, some which typically dwell near the surface and others which live at greater depths. Phytoplankton account for about 90% of the production in the open oceans and form the basis of the food chain there. In addition, there are numerous species that live on or in the ocean bottom or on the continental shelves.

The open oceans are well buffered against extreme temperature changes. It is unlikely that even the most severe nuclear war scenario would result in acute-phase reductions of the sea-surface temperatures by more than a degree or two, and the temperature would not be expected to change in the deep oceans. Thus, little direct impact on ocean species is expected from temperature effects, although there might be indirect effects from changes in ocean currents and circulation. [See below.]

Projected acute-phase reductions in light levels of 95% or more lasting for several weeks could have a more serious impact. It is estimated that this could reduce by a factor of 10 the water volume [and associated nutrients] available for photosynthesis; thus phytoplankton populations could largely disappear. In the most extreme case [light levels at 1% of normal, lasting for months], the phytoplankton standing crop would be insufficient to support the animals that feed off it; however, it is probable that plankton species would not become extinct and would recover after light levels returned to normal.

The loss of phytoplankton could be expected to disrupt the oceanic food chain. Populations of many species dependent on phytoplankton could be affected, especially relatively small organisms without the energy reserves to carry them through long periods without food. It is unclear how intermediate-sized organisms [e.g. fish] would be affected, but many species might be able to find alternative food sources or endure prolonged periods without food. Fish larvae could be more affected than adults.

If larvae are eliminated but not adult populations, the survival of fish species would depend on the longevity and reproductive habits of the adults. There might be little impact on species like cod, which are relatively long-lived and produce large "year classes" [i.e. large numbers of juveniles each year]. In contrast, species like haddock produce a significant year class only every 7 to 10 years, essentially only once in the adult population's life-cycle. In this case, the timing of the post-war disturbances would be important. If they occurred in a year without a large year class of haddock, the species might not be affected at all, but if a large year-class was eliminated, the population could be devastated and the survival of the species could be threatened.

In general, it is not thought that ocean fisheries would be eliminated by post-war effects and, in fact, many species might significantly increase in number after a nuclear war, since a major current cause of mortality is fishing by humans.

It is not expected that acute-phase effects would extend into the chronic phase or that chronic phase conditions would directly cause high levels of stress. However, open ocean fisheries could be vulnerable to indirect effects if ocean currents and circulation patterns were to be altered by post-war disturbances. For example, recent changes in the southern Pacific warm ocean current known as El Niño resulted in the collapse of Peru's entire anchovy industry. Such disturbances of ocean currents have been known to last for years and to interfere with fisheries for years or decades.

Organisms that dwell on or in the ocean bottom away from continents [i.e., the benthos] might be largely unaffected by post-war climatic disturbances. No temperature changes would occur and reductions in light levels would make no difference since these regions are already completely dark. The impact would be limited to indirect effects on their food source, derived from plankton above. If plankton were killed in the acute phase, there might be a brief pulse of increased food, followed by a subsequent decrease. However, benthic organisms could be expected to survive periods without food and probably would be relatively unaffected.

Coastal Ocean Ecosystems

Because of their proximity to land, coastal ocean ecosystems might be more affected by post-war disturbances than the open oceans. They could be more affected by temperature changes and increased storminess, in addition to reductions in light levels. There could also be impacts related to nutrients, sediments and other run-off from land ecosystems.

Phytoplankton populations in coastal waters might be less affected by low light levels than those in the open oceans and their production might not be adversely affected if they could adapt to the unusual timing of an apparent winter.

Coastal ecosystems in tropical waters would be much more sensitive to changes in both light levels and temperature; coral reefs, for example, could be particularly vulnerable to reduced light levels, low temperatures and increased ultraviolet radiation.

Coastal areas such as beaches, mud flats and salt marshes would experience much greater climatic effects than other ocean systems, perhaps including sub-freezing temperatures. Winter conditions in high-latitude coasts already experience freezing conditions. Most mid- and low-latitude coasts do not and their surface-dwelling organisms could be killed, although bottom-burrowing organisms might have some protection.

The cold temperatures could also kill coastal fish species that normally do not experience temperature extremes. It is possible that near-shore fisheries in tropical and sub-tropical waters, and in temperate waters in summer, could be substantially reduced by cold post-war temperatures, especially in bays and other shallow waters that would be most accessible to survivors of a nuclear war. In addition, the eggs or larvae of many fish species live near the surface and would be affected by changes in temperature, ultraviolet radiation, toxic chemicals and other stresses.

Estuaries

Estuaries are salt-water ecosystems that are closely linked to freshwater and land ecosystems [e.g. bays.] It appears that they are both more vulnerable to the after-effects of nuclear war than other marine systems, and of greater importance to humans.

Estuaries could be even more severely stressed by post-war disturbances than near-shore coastal systems; for example, reduced precipitation, not a factor in other marine ecosystems, could result in reduced inputs of fresh water and hence increased salinity.

Low light levels in summer could result in a large-scale elimination of plants in marshes and wetlands and of the animals dependent on them for food. Increases in sediments could literally suffocate bottom-dwelling organisms. Fish species that use wetlands as nurseries could suffer the loss of juveniles.

Some simulation studies suggest that an estuary could recover rapidly after the acute phase with no carryover effects into the chronic phase. However, caution must be used in generalizing from these simulations, which are based on one real site and might not be applicable to other locations, particularly those at different latitudes. Also, the model did not include the possible stresses of nearby nuclear explosions, increased sediments, increased coastal storminess and increased levels of toxic chemicals. Most estuaries in the Northern Hemisphere are closely associated with human populations and, in the aftermath of a nuclear war, many could be expected to experience fallout and increased levels of pollutants in the run-off, resulting from weapons explosions, urban fires and industrial destruction.

SOUTHERN OCEAN AND ANTARCTICA

The Antarctic ice sheet contains about 90% of all the fresh water on Earth and any major long-term change in its mass would have global effects on sea level and more local effects on the atmospheric heat balance and weather systems in the Southern Hemisphere.

The Southern Ocean accounts for about 10% of the global oceanic area. Ice-free land is at a minimum in this part of the world, and most ecosystems are marine-related. Plant life consists almost totally of phytoplankton; in addition, there are marine mammals and large communities of penguins.

Every year, much of the ocean is covered by a large stretch of sea ice, which reaches its maximum in October. Changes in the normal extent or duration of the sea ice would have a major effect on marine ecosystems. If substantial temperature effects and low light levels were to occur in summer, sea ice growth in the fall might begin earlier and extend farther north than normal. This could lead to a minimal summer season the following year and further enhance winter conditions. If the climatic effects occurred in winter, the impact would likely be minimal.

Prolonging the sea ice cover might result in the loss of breeding grounds for Antarctic penguin and seal populations and a shift in the habitats of other animals such as the albatross. At present, it does not appear that permanent alterations of Southern Ocean and Antarctic ecosystems would occur because many indigenous species are adapted to low light and temperature levels and populations are long-lived and widely spread. Some, like plankton, would recover rapidly from any effects that did occur.

ECOSYSTEM RESPONSES TO OTHER NUCLEAR WAR EFFECTS

Ultraviolet Radiation

For the mid-range nuclear war scenarios discussed in Chapter 2, it has been estimated that ozone depletions could reach as high as 20% to 30%, with the maximum effect being reached within six months to a year. This layer protects the Earth's surface from harmful UV-B radiation [see Chapter 6]. Such a depletion in stratospheric ozone could therefore lead to UV-B increases of some 40% to 100% at the surface. In some Northern Hemisphere mid-latitude regions immediately after a war, it is possible that "ozone holes"—characterized by losses of perhaps 70% of ozone over continent-sized areas—could last for up to several days.

Significantly increased levels of UV-B could have severe biological consequences. Most of the known biological effects of UV-B are damaging; for instance, the genetic molecule, DNA, and the protein molecules that form the building blocks of living tissue can be damaged by absorbing ultraviolet radiation. Organisms have developed a variety of defense and repair mechanisms to protect themselves from the UV-B they are normally exposed to, but it is unlikely that these mechanisms would suffice under the large increases in UV-B radiation that might follow a nuclear war.

Land plants have evolved to expose much of their tissue to sunlight in order to maximize photosynthesis. UV-B radiation can have an effect on, for example, leaves and pigments, plant growth and metabolism, fruit growth and yield, and pollen germination. Studies indicate that increased UV-B could cause significant yield reductions in several important crop species [e.g., soybean]. Plants subjected to weeks or months of darkness or very low sunlight levels after a nuclear war would be even more sensitive to UV-B and for a period of time after the war, the ratio of UV-B to the wavelengths of sunlight needed for photosynthesis would be much greater than normal.

Different plant species—and even plants within the same species—would vary in their resistance to increased UV-B. This could cause changes in the competitive balance of various species within plant communities, posing a considerable risk to both agriculture and natural ecosystems. For example, in an agricultural system, increased UV-B might alter the competitive balance of crops and weeds; an increase in weed competitiveness could reduce crop yields or quality or alter their sensitivity to pests or diseases. The species balance in natural ecosystems might also be significantly altered.

In marine ecosystems, the depth to which UV-B radiation penetrates the water is of key importance in assessing its impact on aquatic organisms. For example, it is thought that phytoplankton might move downward in the water to reduce their exposure to increased levels of UV-B. This would also reduce their exposure to sunlight needed for photosynthesis and, in some very productive areas [e.g. high latitude oceans], the level of this photosynthetically active radiation is the major limiting factor on plankton productivity. Phytoplankton are unlikely to move to deeper, lower-light regimes to escape higher UV if this reduces photosynthesis. It has been calcuated that the increase in UV-B resulting from a 25% reduction in ozone could cause a decrease in productivity of about 35% near the surface of the ocean.

Farther down in the water, the effect would not be as great and loss from the total productive zone would be about 9%.

It is possible there would be a shift in the mix of species in marine ecosystems as well. These could affect the value of marine fisheries as sources of food for humans. Zooplankton, small aquatic animals including the egg or larval stage of some important food species, could be seriously affected. These plankton live very near the surface and it appears that current UV-B levels are already near the threshold they can tolerate without being seriously damaged or killed; above these levels, the repair capability is not able to overcome the UV-B damage. The larvae of food fish, which are also near-surface dwellers, could experience similar problems.

Among humans, the consequences of increased UV-B exposure include sunburn, eye diseases, skin diseases including skin cancer, and changes in the body's immune system. There is good evidence for a link between increased UV-B and non-melanoma skin cancer, which, although it can be effectively treated with appropriate medical help, still kills thousands of people a year. UV-B may also be one of the causes of the more lethal malignant melanoma. Finally, UV-B has been shown to alter the immune system. In experiments with laboratory mice, exposure to UV-B reduced the mice's ability to reject tumors. In fact, the UV-B doses that resulted in this impairment of the immune system were much smaller than those needed to cause the tumors themselves. There is concern that such changes in the immune system in humans might contribute to the development of malignant melanoma.

Atmospheric Gaseous Pollutants

In the aftermath of a large-scale nuclear war, large amounts of pollution would be generated in both gaseous and particulate form [see Chapter 6]. It is assumed that humans, agricultural crops, livestock and natural ecosystems in localized areas of the Northern Hemisphere, especially in the mid-latitudes, would be exposed not only to elevated levels of individual pollutants, but to interacting mixtures of contaminants—and under circumstances when they would likely already be stressed by climatic changes, radiation effects and psychological trauma.

A variety of toxic pollutants released in the aftermath of a nuclear war—e.g. nitrogen oxides [NO_x], sulphur dioxide, ozone—could cause serious damage to plants. Individually, sulphur dioxide and ozone are both more damaging than NO_x. It is estimated that ozone is already responsible for significant crop losses in eastern North America and it has been suggested that ozone may be a factor in the decline of forests in eastern North America and central Europe. At high concentrations, sulphur dioxide is known to cause extensive death in forests. Many species of plants would experience severe biological damage after exposure to the local levels of sulphur dioxide that may occur where there is burning of high sulphur sources [e.g., crude oil and coal stores]. The occurrence of acidic fog in low lying areas could add to vegetation damage, while a build-up of carbon monoxide from smouldering fires could be hazardous to people and animals in low lying areas and depressions where they may be sheltering.

Studies suggest that combinations of gaseous pollutants can cause more severe effects than the sum of their individual contributions [this is called *synergism*]. In some plants, the combination of sulphur dioxide and NO_x can cause leaf injury in several agricultural species at concentrations which would not individually be harmful. As another example, tree growth can be suppressed by the combined effects of NO_x and ozone. Maximum damage to some plants occurs when all three gases are combined.

Assessing the potential ecological effects of these pollutants after a nuclear war is difficult because it is difficult to predict the concentrations that might occur and there are gaps in the scientific knowledge about their combined effects. In temperate ecosystems, it is likely that these effects would be more severe if they occur in summer [growing season] than in winter [dormant season]. Some studies suggest that, at lowered temperatures, gaseous pollutants would have lesser effects on crops and ecosystems than would be the case otherwise; however, the low temperatures are themselves harmful to plants, as discussed previously.

Despite the uncertainties, it would appear that gaseous pollutants could represent a significant harmful additional stress on biological systems in the post-war environment.

Ionizing Radiation

Chapter 7 presented assessments of the local and global fallout that might be expected from a large scale nuclear war. Once the radioactive particles are deposited, their behavior and fate depend a great deal on where they are deposited, on their solubility in water and on their chemical properties. The more soluble particles move quickly through both terrestrial and aquatic food chains; other particles are found in soils and sediments and on the leaves of plants.

In the immediate aftermath of a nuclear war, external doses—i.e., exposure of the whole body of plants, animals and humans to gamma rays—would dominate. Over time, however, internal doses from ingestion or inhalation of radioactive particles in food, water and air would come to dominate.

In natural ecosystems, plants would be the first to intercept falling radioactive particles. Some plants—e.g., lichens and mosses—are particularly effective in removing airborne radioactive particles. Finer particles would not be readily washed or blown off leaves, so crops with edible leaves, especially in areas of local fallout, would have to be washed before being eaten to reduce contamination. Waxy fruits and leaves, such as tomatoes or apples, are easily cleaned. For grazing animals, such as cows, sheep, horses or goats, this contamination of foliage and of surface soils would cause substantial internal doses in local fallout areas. Contamination of milk or meat would result.

Some radiation may be absorbed by plant tissues and moved to the roots. It appears, however, that much of the radioactivity is retained aboveground until foliage dies and falls to the ground. This organic litter can become a major source of radioactivity taken up by the roots of other plants. Many plants retain their leaves only for a single growing season, while others retain them for several years or longer. This factor influences the rate at which radioactivity would be cycled through the microorganisms, plants and animals in a natural ecosystem.

Plant species vary a great deal in their susceptibility to radiation. For example, research suggests that coniferous trees are more sensitive than deciduous trees and that cereal crops and native grasses are relatively radiation-tolerant. Non-flowering plants seem to be more tolerant than flowering plants. Radiation-tolerance is greatly enhanced by dryness and dormancy; dormant plants in the dry season are markedly more tolerant than actively growing plants in the wet season. [There is a striking parallel here with cold- and drought-resistance, discussed previously.] Dry, dormant seeds appear to be highly radiation tolerant and the presence of seed banks may be important for recovery in soils that accumulate radioactive fallout.

Among animals, mammals appear to be the most sensitive to radiation and microorganisms the most resistant, although there are, of course, variations among species and individuals, including humans. The amount of radiation damage experienced can be affected by environmental factors, such as pollution and adverse weather conditions.

Insects appear to be more radiation-tolerant than vertebrates. Studies indicate that reptiles are more sensitive than insects and that birds—including those used as human food sources [e.g., chickens, ducks, geese etc.]—appear to be very susceptible. Bird populations in the outer fringes of local fallout areas could be affected, but global fallout doses should be low enough to have no significant effects.[3] As with individuals and species, ecosystems would vary in susceptibility to radiation. Among forests, for example, coniferous forests are likely to be most injured and tropical evergreen forests least injured, with temperate mixed and deciduous forests showing an intermediate response. It is likely that only plants experiencing high levels of local fallout would be seriously injured by radiation alone, although a vari-

---

[3] These discussions apply to external gamma radiation. Although it is less penetrating, beta radiation would not necessarily be unimportant biologically. Studies indicate that it could affect flowering plants and small or thin organisms. It might also damage mouth parts of animals that graze on beta-contaminated plants.

ety of other factors [e.g., age, physiological condition, genetic makeup, etc.] can affect susceptibility. Predicting the extent to which ecosystems might recover from radiation damage is problematical. The amount and extent of damage after a major nuclear war would be unprecedented, making extrapolation from existing knowledge difficult. In general, it would appear that some plants and plant communities could recover from radiation damage as long as some tissues were left intact and soil and climatic conditions permitted plant growth following the stress. The high level of radioactivity in soils subjected to local fallout would affect the ability of land to grow crops with acceptable levels of radioactive contamination. In general, the rate and pattern of ecosystem recovery after a major nuclear war would depend on a great many variables and thus cannot be predicted in detail.

Crop species vary in their ability accumulate radioactivity. Root crops are prone to take up radioactive particles from the soil, while leafy crops accumulate radioactivity through openings in the leaves. Waxy fruits [e.g., apples,cucumbers, tomatoes] tend to be low accumulators. Animal meat and dairy products [especially milk] can also be sources of contamination.

For humans and higher animals, one of the more serious long-term postwar problems could result from the "biomagnification" of radiation in the food chain. As radioactive particles cycle through an ecosystem, they can be greatly concentrated before being ingested by humans. This can cause levels several hundred to several thousand times the levels of radiation in water or in green plants. In terrestrial ecosystems, food chains are relatively short and the radiation is not likely to be magnified more than a few hundred to a few thousand times. In aquatic ecosystems, however, the concentration could be much greater. Freshwater ecosystems—lakes, rivers, streams and groundwater—could be contaminated directly by rain, snow or dry particles, by surface run-off and by leaching from soil. Surface waters could be polluted for a prolonged period. In the short-term, groundwater contamination would likely be minimal since most of the radioactivity would remain in the overlying soils. However, acid precipitation could increase the leaching of radioactive particles from the soils and groundwater, once contaminated, would likely remain so for much longer periods than streams and lakes. Biomagnification of radioactive particles would occur through aquatic plants and animals; this process could be affected by a number of environmental factors, such as water chemistry, mineral content, interactions between water and sediments and others. In some cases, the biomagnification would be significantly greater in freshwater than in salt water ecosystems.

In general, marine ecosystems are likely to be somewhat less sensitive to radiation than freshwater systems because of greater dilution. Radioactive particles could be mixed through the ocean surface layer [5–200 meters] within days or weeks. They could be moved up and down by upwelling and sinking of water, by the day-night movement of plankton and the sinking of faecal pellets from organisms living in the surface layer. Movement into the deep ocean, however, could take decades, even centuries. The bioconcentration of radioactivity is much less in marine ecosystems than in freshwater ecosystems. However, some marine animals—molluscs and crustaceans, for example—are capable of considerable bioconcentration. In the aftermath of a nuclear war, these shellfish might be contaminated to a dangerous level, since they are harvested in shallow waters, which could contain high levels of radioactivity from surface run-off.

Internal Doses in Humans

In the early period after a nuclear war, humans would receive internal radiation doses primarily from inhaling contaminated air and ingesting radioactive particles on the surface of foods. Initially, it would be possible to reduce this source by carefully washing the food. After radioactivity had become incorporated into the food itself, there would be no easy means of decontamination and humans would be subject to chronic internal radiation doses from contaminated food and water. This radioactivity could be particularly harmful to lungs, the gastrointestinal tract and various organs and could result in cancer, reduced immunity to disease and increased weakness to other stresses.

Groundwater is not likely to become significantly contaminated on a large scale, but if it were, it would remain hazardous for decades and could not be easily decontaminated in a post-war environment. Surface waters of lakes, streams or rivers are likely to experience high inputs of radioactive material due to erosion or drainage from land. Such waters may be hazardous to people and to domestic animals in high fallout areas.

HUMAN CARRYING CAPACITY OF NATURAL ECOSYSTEMS

In the next Chapter, the human impact of the potential post-war loss of food production and imports is examined. Therefore, it is useful to consider here whether natural ecosystems could support the human population if the global agricultural system were to be totally eliminated in the aftermath of a large-scale nuclear war. Such a total collapse is not being predicted, but this situation, if it were to occur, would represent the outer bounds of human vulnerability to the loss of the global food support base. It is in this light that the potential complete reliance on natural ecosystems for food is examined.

The current human population of 5000 million people can only be maintained with the food production and distribution systems of modern agriculture. As outlined in Chapter 9, agricultural productivity could be severely impaired after a major nuclear war, resulting in a potential food crisis of unprecedented magnitude. It is likely that humans would place increasing demands on natural ecosystems, but even if these ecosystems were not severely stressed by climate changes and other effects of a nuclear war, they could maintain only a very small fraction of the current world population.

The size or density of human populations that could be indefinitely sustained by food from natural, non-agricultural regions is called human carrying capacity. Without agricultural production, food would be limited to the small fraction of natural biota that could be harvested and digested by humans. Compared with agricultural systems, natural systems produce relatively small amounts of usable food energy. Their present human carrying capacity is estimated to be probably below 1% of the world population and, in a post-war environment, the increased human demand on their resources would likely coincide with historically unprecedented environmental stresses that could further reduce their already very low carrying capacity. Thus, sole reliance on natural ecosystems could result in the nearly total elimination of the global human population.

The historical record shows that the complete absence of agriculture is associated with very low densities of human population. In these circumstances the highest population densities are possible in coastal or floodplain regions, where both land and aquatic resources could be harvested. However, oceans could not be fully exploited by survivors of a nuclear war. Forests contain mostly undigestible fiber [wood] and unpalatable or toxic foliage.

Great skill and knowledge would be required to exploit natural ecosystems near their human carrying capacity. Even if these skills were applied, human populations could be maintained only at reduced levels, comparable to those of hunter-gatherer societies. Animal populations, on which humans would also rely for food, would be similarly hard-pressed to maintain their numbers. Even under the most favorable circumstances only small numbers of humans could be supported by harvesting wild animals.

In summary, it would appear that only very small fractions of the current human population could be maintained solely on natural ecosystems. It is estimated that population levels could be reduced by a factor of 100 or more. Again, it is not being predicted that this will happen, but it is suggested that the current levels of world population would be extremely vulnerable to a global-scale disruption in agriculture.

CHAPTER 11
The Human Response

It has been said that one of the major difficulties in studying the environmental and societal after-effects of a large-scale nuclear war is that there are no historical precedents. And yet, we are not totally without real-world evidence from which some extrapolations can be made—evidence from the nuclear bombings of Hiroshima and Nagasaki; the conventional bombings of Tokyo, Dresden and Hamburg; the effects of large natural disasters such as earthquakes, floods, epidemics, famines etc. The evidence from these events can be instructional, but it is important to be aware that only limited extrapolations can be made from them to a full-scale nuclear war, which would, in all probability, unleash a level of environmental, social, economic and human devastation unprecedented in magnitude and extent.

To gain some understanding of the differences involved, computer simulations were done in which a 1-megaton modern weapon [which is about 65 times the size of the bomb dropped in 1945] was assumed to be exploded in Hiroshima of 1980, both as an air burst and as a ground burst. In the 1945 case, more than a third of the population of 320,000 were killed and about a quarter were injured. In the simulation of an air burst, more than half the nearly 1.1 million population were killed and another 40% of the remaining people were injured. [In the ground burst simulation, the percentage of fatalities was lower than in the air burst simulation, in which blast and thermal pulse would affect a larger area, but higher than that for the 1945 bomb. Injuries in the ground burst case were a lower percentage than for either of the other cases. Unlike the situation in 1945, initial nuclear radiation would not contribute to deaths since anyone close enough to receive a lethal dose of such radiation would be close enough to be killed outright by blast or thermal pulse.]

This scenario, however, again assumed a single detonation in one city. Multiple bursts over many cities—a full-scale nuclear attack—could produce qualitatively different effects, for which the 1945 bombing of Hiroshima and Nagasaki provides no precedent. They include:

- *Direct effects from other detonations in the same city.* Doses from local fallout would be additive, but the blast wave from one explosion could make the area more vulnerable to the effects of blast from a second;

- *Additional local fallout* from surface explosions in other cities upwind;

- *Lack of outside assistance* in providing food, medical help and supplies and social order. This was the key to the long-term recovery of Hiroshima and Nagasaki; the redevelopment they experienced could not be repeated in the aftermath of a large-scale nuclear war.

- *Disruption of communications* from the initial electromagnetic pulse and damaged equipment would hamper recovery efforts;

- *Psychological effects*: Survivors would know [or assume] that most of the world's civilization had been destroyed and that there would be little possibility of outside help. Some would also have heard about the potential for extreme climatic disturbances following the war. This knowledge could have a psychological impact beyond anything experienced in Hiroshima and Nagasaki.

- *Economic effects*, including the disruption of food exports, could cause serious food shortages and starvation in a short period of time. The loss of fossil fuel energy supplies and products derived from them would affect post-war agricultural productivity. Survivors would be limited in their ability to migrate, to manufacture essential goods, to obtain uncontaminated water and to overcome a host of other problems.

- *The direct climate effects* could subject survivors to subfreezing temperatures and near darkness within a few days of the explosions. Many of those injured and left homeless would die from exposure.

- *The indirect effect of climate changes* could result in reduced agricultural productivity that would prolong food shortages; many could die of starvation in the chronic phase. Finding water would be difficult if there were long-term reductions in precipitation.

- *Other environmental effects* could include the outbreak of disease-bearing pests; increased air and water pollution; radioactive contamination of food and water; flooding from melting ice and snow and also because of lost vegetation cover; increased levels of UV-B; increased internal doses of nuclear radiation due to eating, drinking and inhalation.

It has been estimated in various studies that several hundred millions of people could die from the direct effects of a major nuclear war. These calculations are, of course, dependent on the war scenario assumed; in one severe-case analysis involving targeting of most cities of the world, the World Health Organization's estimates went as high as 1,100 million fatalities.

Most of these casualties would occur in combatant nations and their close neighbors, probably almost entirely in the Northern Hemisphere and mostly in the mid-latitudes. It has been estimated that targeted countries could lose between half and three-quarters of their populations to direct effects and that the social and economic support systems in these countries could collapse completely.

The direct effects of nuclear war—blast, thermal radiation, ionizing nuclear radiation and fires—would be fairly localized in the vicinity of the explosions, as described in Chapter 3. Local fallout would occur over large regions and would be almost solely responsible for the radiation-induced fatalities of the war, largely within the boundaries of targeted nations, except for parts of Europe which might receive substantial doses of local fallout even if not targeted. Global fallout, which would occur in non-combatant nations, is not projected to reach high enough levels to cause widespread fatalities from acute radiation exposure.

The number of fatalities in the immediate post-war period would be affected by the likely collapse of societal systems, particularly the loss of medical personnel and facilities, as the experience of Hiroshima and Nagasaki suggests. In Hiroshima, a well-prepared medical rescue system was rendered completely useless. Of the 298 doctors in the rescue squads, 270 were killed immediately by the bomb and 8 to 9 out of every 10 nurses were also killed. The Nagasaki Medical University and three other hospitals were demolished and burned and almost all the medical staff died instantly or within a few days.

It should also be remembered that, for Hiroshima and Nagasaki, there was an "outside" world capable of providing some assistance, although the sheer numbers needing help [about 150,000 refugees in the case of Hiroshima] initially overtaxed the capabilities of surrounding towns and villages. In the two months after the bombing, however, there was an influx of some 3000 medical personnel. The war scenarios presented in Chapter 2, some of which assume much more intense bombing of cities with much larger-yield weapons, imply that, for many regions in targeted countries, there would be no outside help, since there would be virtually no "outside." Non-combatant nations dependent on imports from Northern Hemisphere industrial nations might find themselves similarly without the possibility of outside assistance if faced with post-war food or energy shortages.

The number of survivors of the direct effects who would die directly from exposure to low temperatures in the acute phase after the war would be determined by the availability of warm clothing, shelter and energy supplies. It is not thought that the number of deaths from this cause would be large, although people in the tropics might be susceptible to even relatively small temperature decreases because of their lack of adaptation to cold and the unavailability of protective clothing and energy supplies.

However, the indirect effects of a nuclear war—particularly those related to food availability—are likely to have a much more serious human impact.

As the following analyses will show, the impact on the remaining global population from the loss of one growing season in response to climatic and other disturbances after a nuclear war would, *at a minimum*, far exceed the direct effects from nuclear explosions. Although the number of fatalities that would result from the direct effects would depend on the war scenario adopted, it is impossible to target the weapons to produce the number of direct casualties that are estimated to be possible from food shortages in the chronic phase after the war—perhaps as many as 1000 to 4000 million people.

The food shortages would result not only from the potential losses in agricultural production discussed in Chapter 9, but possible reductions in or total loss of food imports, human subsidies for agriculture and indigenous food stores. The need for food imports and food stores, particularly in the first year after the war, could vary considerably, depending on the time of year in which the war started, i.e., the extent to which nuclear war-induced climatic stresses interfered with crop harvests.

LOSS OF FOOD IMPORTS

It is likely that food exports from countries that currently export grains would be terminated after a nuclear war because the major exporters would likely be combatant countries and because the world's economic, trade and food distribution systems would be disrupted. The vulnerability of various regions to the loss of gross imports can be assessed by examining the fraction of total food resources now made up by imports: Africa, 23%; Asia, 11%; South America, 18%; Europe, 20%; North America, 13%. [On a country-by-country basis, the range is much wider—from less than 1% to 99%.]

It is difficult to translate the loss of food imports directly into reductions in the human population, because other factors must be taken into account [e.g. the diversion for food production of land now used for export and non-food crops, the diversion of food from animals to humans, changes in human diet etc]. On the other hand, the historical evidence from famines indicates that human populations are extremely sensitive to the loss of even relatively small fractions of total food resources.

This vulnerability, of course, varies widely among countries. Gross imports in Australia, for example, are less than 0.05% of its grain production and it would probably suffer no effects from the loss of imports. Japan, on the other hand, would be very vulnerable; one analysis suggests that, if food imports ceased, only about half the population could be supported on indigenous food production if that were to continue at current levels. Even in the absence of climatic or other physical effects from the war, widespread fatalities from starvation could be expected under these circumstances.

THE LOSS OF HUMAN SUBSIDIES TO AGRICULTURE

The potential climatic after effects of a large-scale nuclear war could dramatically reduce and even eliminate agricultural productivity in many regions of the world for perhaps several years after the war. In subsequent years, the task of rebuilding the agricultural system in combatant nations and of increasing production levels in non-combatant nations could continue to be affected by climatic factors. These difficulties could be further compounded by the lack of transportation facilities, agricultural machinery, energy supplies, fertilizers, pesticides, hybrid seeds and the other human subsidies to agriculture that permit production levels capable of sustaining the current world population.

Indeed, even *without* climate disturbances, global agricultural productivity would be highly vulnerable to the loss of technological and energy subsidies that would occur in the aftermath of a nuclear war. The developed nations of the Northern Hemisphere, the major source of these subsidies, would suffer the most severe direct damage from nuclear explosions, and industrial activity in these countries would be seriously disrupted. Moreover, the entire transportation and communications infrastructure of international trade would be disrupted, so that many non-combatant nations dependent on imports could also be left without essential subsidies.

Because agricultural systems in developed countries like Canada, the U.S., the USSR, Australia and the United Kingdom depend on high energy inputs to maintain productivity, they would be highly vulnerable to disruption after a nuclear war. Agricultural systems in the developing nations are less dependent on energy inputs and therefore would be less susceptible to major loss of productivity; however, they are becoming more dependent on subsidies.

The relative vulnerability of various agricultural systems can be assessed by examining their relative dependence on the various types of human subsidies now put into agriculture.

Fossil Fuel Subsidies

During the last century, virtually all agricultural advancement depended primarily on the direct or indirect use of fossil fuels. Direct uses include running farm machinery for planting, cultivation and harvesting; indirect uses include the production of fertilizers and pesticides. Within the past half-century, the use of these subsidies has increased agricultural production 3- to 4-fold in developed countries and about 2-fold in developing countries.

Fuel supplies for running agricultural machinery could be severely interrupted after a major nuclear war, both in combatant and non-combatant nations dependent on imports. The continuing availability of machinery and replacement parts could be a problem, since energy would be required for the manufacture, repair and distribution of such equipment.

The potential loss of agricultural machinery could have a major impact on post-war agricultural productivity, given the dramatic reduction in labor inputs that have been achieved through the introduction of engine power. For example, the direct and indirect labor input in U.S. maize production is about 2% of the input required to produce maize by hand.

In many tropical developing countries, draught animals are used for agricultural production. If these animals were to be eaten for food in the acute phase after a nuclear war, production could be reduced to that which could be accomplished by hand.

Irrigation is also a critical requirement in many tropical regions where important food crops are grown in the dry season. Power shortages and lack of maintenance of irrigation facilities could cause massive crop losses in these areas.

Fossil fuels are also used to produce fertilizers and pesticides. In many regions, particularly in developed countries, the high levels of productivity in maize, wheat and rice have been achieved only by heavy applications of these inputs. Energy is also required for drying grain and for irrigation.

Major agricultural production systems in developing countries also depend heavily on fertilizers. Nearly a third of total world fertilizer use occurs there. China, for example, uses more per hectare than many developed countries. Rice yields in the developing world have been doubled since 1950, largely because of intensive inputs of fertilizers. In India, more energy is used in one season in the form of nitrogen fertilizer for paddy rice than in the form of human labor and bullock power combined. In Central America, fertilizer use accounts for more than half the energy subsidies.

The loss of nitrogen fertilizers after a nuclear war would not be immediately catastrophic in regions with fertile soils or with a long history of prior fertilization. Generally, however, unfertilized soils would provide inadequate nutrients for plant growth.

In conclusion, it would appear that the agricultural technologies in developed countries would be disproportionately affected after a major nuclear war because of their sophistication and dependence on fossil fuels. Even in the absence of severe climatic disturbances, the agricultural systems of the developed world would be highly vulnerable.

In regions of the world that are not directly devastated by the war, agricultural production might or might not continue to receive direct and indirect energy subsidies. Some areas could be nearly self-sufficient in energy. In other regions which are not now heavily dependent on subsidies, agricultural practice could revert to simpler methods but with some loss of productivity. The latter two types of regions, likely to be mostly in non-combatant nations, would be least vulnerable to the loss of human subsidies to agriculture.

Alterations in the climate could shift crop ranges, so that the normal crops could not survive and different crops would have to be planted. Even if seed or rootstock could be found, farmers might not have the cultural knowledge to manage the unfamiliar crops successfully. [Indeed, many of those attempting farming might be former urban dwellers with little or no farming expertise.]

The lack of machinery and other energy subsidies might mean that different, perhaps more primitive, farming techniques must be practiced. The risk of crop failures might have to be hedged by increased use of intercropping [growing different crops with different vulnerabilities at the same time]. Land not currently used for food production might have to be converted for this purpose [a strategy that might be successfully employed in many tropical countries which currently use a sizeable fraction of their arable land for non-food and export crops]. Thus, while productive agriculture might be technically possible in many areas, its success would depend on a flexible response to the difficult task of matching the correct crops and farming practices to altered but unpredictable climate conditions.

FOOD AVAILABILITY AFTER NUCLEAR WAR

How much of the human population could be supported by the global agricultural and food distribution system after a large-scale nuclear war?

Food shortages would be a major problem facing survivors of a large-scale nuclear war, especially those in non-combatant nations who suffer far less from direct effects of a nuclear war. These shortages could result from a complex combination of factors—climatic stress on agricultural systems; the interruption of international trade, with a subsequent loss of food shipments and human subsidies; lack of fuel; radioactive contamination; uncontrollable fires; the destruction of national and international facilities for storing, transporting and distributing food; and—perhaps not the least important—potential widespread social unrest.

It is possible that food production in most of the Northern Hemisphere and much of the Southern Hemisphere could be virtually eliminated for at least a year after the war. Under these circumstances, the amount of food held in storage would become a critical factor controlling human survival in the acute phase. Even with minor climatic effects, many countries could be expected to suffer severe food shortages due to lack of food imports and energy.

An analysis has been made of 15 countries that make up about 63% of the global population and are responsible for about 43% of world agriculture. These countries represent a wide spectrum in terms of population levels, agricultural productivity and economic and social structures. They are also representative of different regions of the world that would experience varying climatic effects in the aftermath of a major nuclear war. They include Northern Hemisphere industrial nations [including combatants]; Southern Hemisphere tropical and mid-latitude countries [including non-combatants]; food-exporting and food-importing countries; developed and developing countries; island nations; and the most populated countries on Earth.

An additional 120 countries were analyzed in less detail. The results of these studies indicate that *food problems could be the single most significant contributor to human mortality following a nuclear war.* Thus, virtually the entire human population of the Earth—not just those in combatant nations—is vulnerable to the large-scale use of nuclear weapons.

The amount of food stored varies greatly nationally and regionally; for example, countries that are major exporters of grain generally have a higher level of "carryovers"—i.e., the amount of food left over from one year's production at the time of the next year's harvest. Food stores also fluctuate during the year, depending on the size and timing of harvests and on imports and exports shipments. Thus, the timing of a nuclear war—e.g. before a harvest or after—could have a great impact on food stores.

The following analysis of potential food shortages in various countries around the world involves a number of assumptions:

- A minimum of 2000 kcal/person/day [of which about 75% is composed of cereals] would be necessary to sustain life for an extended period. This would represent a significant change in the normal diet of most people in developed industrial countries.

- Food stores and population are destroyed in equal proportions in combatant countries. The assumed levels of destruction ranged from 25 to 75% of pre-war levels. However, it is unknown how vulnerable food stores would actually be to destruction; this is a subject that requires further research.

- Food distribution within a country is optimal, in that the maximum number of people survive, given the minimum diet specified. This implies full and equal access for all who survive. However, the evidence from famines demonstrates that food is rarely distributed optimally, so this assumption leads to a considerable over-estimate.

- No food is used by people who cannot be supported for a full year. In other words, the people who would *eventually* die from starvation [i.e. the number of people that exceeds the one-year food supply capacity] would die instantly at the beginning of the post-war period, so that they would consume no food which would otherwise go to the long-term survivors. This, again, is an assumption biased in favor of maximum survival of humans.

- No animals are fed with grain.

The human support capacity of food stores—i.e. how long the food could support the full surviving population—was calculated for the acute and chronic phases after a war, assuming no agricultural production or food imports for one year. Since many of the assumptions are optimistic, the resulting calculations should be considered upper limits of human survival. More realistic assumptions would lead to lower estimates. Moreover, this analysis involves no speculation about the impact on food availability and distribution of human behavior in the face of food and resource shortages.

There is no way of knowing the combined impact of other factors that seem likely to ensue: food hoarding and maldistribution, social conflicts, vitamin and protein deficiencies, increased food requirements associated with increased manual labor, food spoilage and contamination and less-than-perfect allocation to only those who would eventually survive. All of these factors could result in a considerable reduction in the estimated numbers of people who could survive a nuclear war in the long term.

Following are some of the results from the sample of 15 countries for which a detailed analysis was done:

Australia

Australia is a Southern Hemisphere net food exporter. Unless it was targeted in a nuclear war, it would not suffer direct effects, but Australian agriculture could be affected by potential post-war reductions in temperature and especially precipitation. However, the country has large grain stores and large numbers of sheep and cattle so even complete elimination of agricultural production for one year would not necessarily lead to acute food shortages. Pastoral agriculture would be more resistant to climatic disturbances than other types of agriculture. It was concluded that there would potentially be enough food in Australia to carry the entire surviving population through the one-year acute phase with no agricultural production, regardless of the timing of the war.

Brazil

Brazil spans equatorial and Southern Hemisphere latitudes and is very diverse climatologically. It is not considered likely to be directly targeted in a nuclear war, but nevertheless could suffer severe after effects because its agriculture is vulnerable to even brief episodes of cold temperatures. A large fraction of the current agricultural production is needed to support Brazil's present population; therefore, the country is unlikely to have sufficient stored food to support its population during a one-year acute phase following the war if agricultural production were severely reduced or eliminated.

In the chronic phase, recovery would be hindered by the lack of imports of energy and other subsidies and possibly by continuing chronic-phase reductions in temperature and precipitation. Some compensation would be possible by growing food crops on land now used for non-food and export crops and possibly by shifting production to areas where the altered climatic conditions are suitable. However, agricultural impairment seems likely, given the dependence on imports, and the addition of climatic stresss could reduce production below the levels needed to support the current population.

Canada

Canada is a major food exporter, but is also a Northern Hemisphere industrialized country likely to suffer severe direct consequences in a nuclear war. In the acute stage, its agriculture would likely be greatly affected by post-war climatic disturbances. As was discussed in Chapter 9, reductions of even a few degrees C in average growing season temperatures could result in the loss of the wheat crop. Even relatively mild climatic changes could virtually eliminate Canadian agricultural production in the first growing season and, in the worst climatic circumstances, possibly for several years.

Nevertheless, Canada need not suffer an acute phase food crisis. Since the ratio of food production to population is large, it is concluded that the surviving Canadian population could be fed on stored food for an extended time, assuming the means to deliver the food.

Despite the strength of the Canadian agricultural system, recovery in the chronic phase could be hampered by the lack of energy and energy-related subsidies if industries and refineries were destroyed in the war. Continuing climate stress could also be a problem. So, although only a small fraction of pre-war agricultural production would be needed to support the Canadian population, it is possible that production could be reduced even below these levels.

Peoples Republic of China

China would probably suffer a food crisis after a major nuclear war even if it were not directly targeted. Its agricultural production and food supply system, which must maintain a high level to support 1000 million people, would be vulnerable to the extreme climatic disturbances projected to be possible for mid-continental, mid-latitude Northern Hemisphere locations— e.g. an average drop in temperature of 10°C to 35°C for several weeks following a summer war, with smaller reductions possible for an extended period. Even temperature reductions at the lower end of this range could produce climatic conditions near the tolerance limits for rice crops in much of China and could eliminate rice production.

If a full annual harvest were in storage at the start of the war, adequate food would be available for the acute stage, but if only carryovers were available, the food stores would support only a small fraction of the Chinese population. In the worst cases, a majority of the Chinese population would die from food shortages.

Continued agricultural production seems possible in the chronic phase if China were not targeted in the war and the climatic effects were not large; otherwise, production levels could be reduced well below the levels needed to support the present population. China is particularly vulnerable to chronic phase climatic effects because of its great dependence on rice, which is a very cold-sensitive plant. In addition, Chinese agriculture, especially rice, is energy and subsidy intensive; the dependence of the rice crop on irrigation is a major vulnerability. Although the country currently produces most of these subsidies itself, disruptions following the war, especially if the country were targeted, might eliminate the industrial capacity needed to supply them.

India

India could suffer consequences as severe as those faced by combatant countries in a nuclear war. The country spans a wide latitude band in the Northern Hemisphere and has a wide range of climates. Some inland regions could experience the more severe projected temperature reductions in the acute phase and, over the longer term, reductions in precipitation associated with failure of the monsoons. These factors could eliminate agricultural production, which would likely create a severe food crisis. India has little capacity to feed its population on stored food and even a full harvest in storage would not sustain the population during the one-year acute phase. In the worst scenarios, most of the Indian population would die from food shortages.

Recovery would be difficult if the chronic phase were characterized by continued reductions in precipitation and temperatures, which could affect the rice crop in much of India. Indian agriculture at present is highly dependent on energy subsidies. With severe climatic disturbances, and possibly targeting, it probably could not compensate for losses of energy imports; this crop yields could be expected to decline.

Japan

Although it is assumed that Japan would be targeted in a major war, and would therefore suffer from direct consequences, the impact on its agriculture would still be devastating even if it were not directly attacked. A food crisis and many deaths would be possible in the acute phase. Japan is a net importer of grain and it would be unlikely that its food stores could support the full population for a year. Even relatively mild climatic stresses could eliminate agriculture production for a year or more, with Japanese rice being particularly sensitive to even the less severe projected temperature decreases.

Agricultural recovery could be difficult in the chronic phase. Since it is downwind from a major continental area that might have temperatures well below normal, Japan might suffer continuing low temperatures that would inhibit agricultural production. In the worst cases, rice production could be eliminated for one or more growing seasons. Without climatic stresses, recovery would be hampered by the lack of subsidy imports for Japan's heavily energy-intensive agricultural system. Because of its import dependence, Japan's current population could not be sustained in the chronic phase even with full agricultural production; the additional impacts of direct effects from nuclear explosions, climatic stresses and energy shortages would devastate Japanese agriculture and the population.

United States

U.S. agriculture would be brought to a stand-still by the direct effects of nuclear explosions and the most severe climatic consequences projected for a large-scale Northern Hemisphere war—average temperature decreases of as much as 15°C–35°C for a few weeks and short-term, large reductions in precipitation. Under such circumstances, grain production would be completely eliminated for at least one year and only a small fraction of animal products and other crops could be maintained.

The exporting of food would cease, which would have an impact on noncombatant nations dependent on food imports. As a massive food producer, however, the U.S. could potentially support its entire surviving population for several years [assuming that between 50 and 150 million were killed in the bombings] but there could be serious problems with food distribution.

Societal disruptions and continued reduced temperature and precipitation levels could hinder agricultural recovery for several years in the chronic phase. The current U.S. agricultural system is heavily energy-intensive and the industrial capacity to produce these subsidies could be considerably reduced after a nuclear war. Although even a small fraction of current agricultural production could support the surviving U.S. population, it is possible that production may fall even below these levels.

Union of Soviet Socialist Republics

Like the U.S., the Soviet Union would suffer substantially from direct effects of nuclear weapons explosions and the most severe projected acute-phase climatic consequences. In addition, much of Soviet agricultural production occurs in northerly latitudes. These areas would be especially vulnerable to post-war climatic disturbances and, as in Canada, Soviet agriculture could be completely eliminated in the first post-war year. Food stores could probably support the surviving population through the first year, provided distribution systems were intact, unless the war occurred while annual food storage levels were at a minimum.

In the chronic phase, the Soviet agricultural system would continue to be vulnerable. Temperature reductions of several degrees and precipitation decreases of up to 50% could continue for an extended time. Even without climatic stresses, agricultural recovery could be impaired by lack of industrial and transportation systems and energy subsidies. With no climatic stresses, Soviet agriculture might be able to support the entire pre-war population but prolonged climatic disturbances could reduce production to a small fraction of current levels.

Calculations from these studies indicate that populations in these countries could be reduced between 0% and 40% after a nuclear war simply from a lack of food imports and energy subsidies. It should be noted that the effects of possible acute or chronic climatic changes and immediate fatalities from nuclear explosions are not included in these calculations but, if they were, they would not necessarily be additional, for the simple reason that you can't kill people twice. In other words, if 20% of the population were killed by the immediate, direct effects of the war, food demand among the survivors would be reduced and perhaps more than 40% of the pre-war population would survive. Similarly, if large numbers of people die from crop failures and food shortages in the acute phase, food demand would be reduced in the chronic phase.

When the effects of potential chronic climatic changes are considered, it is estimated that about 30% to 80% of existing populations among the 15 countries could be maintained over the long term. The higher figure assumes minimal climatic effects but the loss of human subsidies for agriculture.

Extrapolating from these figures to the total global population, it is estimated that about 1000 to 4000 million people could die in the chronic phase. This does not include those who would die from the direct effects of the war or from starvation in the acute phase. The previously stated caveat about effects not being additive would apply to this calculation as well.

LONG-TERM AGRICULTURAL REDEVELOPMENT

A question remains about the potential recovery of natural, agricultural and societal systems after the war. The current state of knowledge does not permit firm estimates of the speed or extent of recovery, but some major factors that would influence redevelopment efforts can be identified.

Physical Factors

While it is believed that climatic extremes would not continue for a long period of time after a nuclear war, long-term persistence of significant climatic changes cannot be ruled out. Survivors could be faced with long-term increases in the risk of crop failure or large losses of crop production resulting from: continued changes in average climatic conditions; more frequent episodes of extreme weather; and continued lack of subsidies.

In addition, the longer-term climate problems could exacerbate the loss of seed supplies for three reasons. If the acute phase occurs prior to crop maturation, the loss of a crop would mean the loss of the seeds derived from it. If crop productivity were reduced during the chronic phase, the fraction of the crop set aside as seed might also be reduced because of the more immediate human demands for food. Finally, seeds used for crops that subsequently failed—say, during an extreme climatic episode during the chronic phase—would be effectively wasted.

Another physical effect that could create long-term problems is damage to the productivity of the soil. Post-war fires and the failure to protect soils during farming could result in widespread erosion, loss of nutrients and possibly desertification in some areas. Soils might also be contaminated for long periods by radiation from the local fallout and by toxic heavy metals from urban fires, which could reduce crop productivity, reduce the area of land safe for cultivation and result in human health effects.

Biological Factors

Interactions between agricultural areas and natural ecosystems could cause potential problems. For example, disturbances to ecosystems could influence the spread of insect and other pests to agricultural areas at a time when the availability of pesticides would have all but disappeared. Even attempts to introduce natural pest predators could be compromised by the large uncertainties in predicting pest outbreaks in the severely damaged ecological systems. Continuing environmental stresses would also increase the probability of plant diseases and spoilage in stored crops.

Information and Technology Factors

The availability of information needed for successful farming is another important factor that would likely be affected by post-war conditions. Decisions must be made regarding the timing of crop cultivation and the suitability of different crops to the areas in which they are planted. Experience is a valuable asset in determining planting dates and seed types, but in a post-war situation, farmers would not be able to rely on information from the immediate past, since the nature of the future climate would be very uncertain. Moreover, it would be difficult to apply knowledge gained in one growing season to the next or in one area to another. Given the likely lack of communications facilities, it would be impossible to produce long-range weather forecasts, which are based on inputs of world-scale weather information. In any event, the post-war atmosphere might be very different from "normal". Finally, extreme weather episodes would not be easy to predict, even if good communications existed.

Planting strategies, such as intercropping, could be employed to reduce the risk of crop losses, but there could be problems with the adequacy of the information base on which such decisions could be made, with distribution of this information and with the availability of seed sources.

Farm machinery would not be widely available and even those machines that remained operational would be limited by the lack of fuel and, eventually, replacement parts. It is likely that non-mechanized farming practices would have to be adopted and draft animals would become very important for tilling, cultivating and harvesting crops. It would take time to breed animals to replace those killed directly in the conflict or eaten for food and there could be problems with a lack of breeding stock. There would also be maldistribution of available animals.

Human and Societal Factors

One of the most important requirements to resume agricultural production would be the availability of seed. The possible widespread destruction of food stores, increased food demand among survivors and lack of know-how by those unfamiliar with agricultural practice could cause the loss of large quantities of seed. Even if seed does survive, the loss of distribution systems may hamper efforts to re-establish agriculture. In addition, seeds obtained might not be suitable for the [possibly altered] local conditions. Attempts to shift to cold- and drought-hardy crops might suffer from the lack of appropriate seed.

It should be remembered, too, that many of the people attempting farming in the aftermath of a nuclear war might lack information, experience and cultural knowledge concerning agricultural practices. Regions in which largely rural populations have only recently moved into cities might fare better than regions which have long been industrialized. People unfamiliar with methods of conservation or who rushed production to meet food demands could overexploit farm land. There could be problems associated with the increasing reliance on human labor for agriculture if people are weakened by lack of food and inadequate nutrition.

It is possible that societal conflict could also reduce agricultural productivity and distribution of food—for example, if wandering groups invaded agricultural areas in an attempt to obtain food by force from those who were farming. On a wider scale, there might be competition for limited resources between combatant and non-combatant nations, especially on a regional basis.

It has been estimated that 1000 to 4000 million human beings might die as a result of starvation in the chronic phase after a nuclear war. Does this mean the loss of all humans on Earth? The best current estimate is that this would not necessarily happen. And it should be remembered that these projections of global-scale losses do not mean that, even in areas where humans would be expected to die, all would suffer the same fate. Small groups—at the level of families or communities—might manage to survive,but it is impossible for thousands of millions of people to do so, lacking an adequate food base.

Estimating the Effects on the Social Structure

Modern cities are sophisticated, highly integrated technical-industrial complexes whose functioning would be seriously impaired, if not totally destroyed, by a large-scale nuclear war. In the case of cities that were not totally destroyed, questions can be raised about the flexibility of such systems—and the ability of humans to manage them—in the face of the stresses that could occur after the war. Studies of human response to other disasters indicate that relief, rescue, emergency aid, evacuation and reconstruction are accomplished by societies that maintain their social organization, are able to accept evacuees and can provide some outside assistance.

As we have seen, these situations may not apply in the case of a large-scale nuclear war. The widespread destruction might preclude evacuation and outside aid probably would be unobtainable. Fire-fighting would likely be impossible. Reconstruction would be hampered in areas contaminated by radiation. Local and national governments might no longer be able to provide security, transportation and distribution of food. All problems could be exacerbated by the lack of communications.

No society has ever been subjected to this kind and magnitude of disaster and therefore, predicting social responses on the basis of past catastrophes is highly speculative.

Even more problematical is the societal response in the face of the climatic disturbances that might follow a nuclear war. In addition to the problems cited above, there would be the physical and biological impacts discussed throughout this book—the reductions in temperature, sunlight and precipitation and the accompanying impacts on ecosystems and agricultural production. This could stress the human social fabric in unique and fundamental ways.

Epilogue

There is compelling reason to believe that a large-scale nuclear war could be not only war among combatant countries, but war on all the Earth's nations and peoples, and upon the global environment and biosphere as well. Despite the uncertainties outlined in the preceding chapters, the scientific evidence strongly suggests that detonation of a large portion of the world's nuclear arsenals—especially if this results in large-scale urban fires—might lead to climatic and biological consequences that could prove devastating to much of the Earth's population. These findings give credence to the view that there could be no winners in a major nuclear war, nor could opting out of the conflict insulate nations from the consequence of a such a war.

This much seems evident on the basis of scientific research done to date. As the earlier chapters have documented, some general characteristics of the post-nuclear war world also seem quite likely—for example, that there could be serious climatic cooling in many regions of the Earth, particularly in areas where nuclear bombs demolish cities, and that such cooling could have a devastating effect on world agriculture. This implies that, for much of the Earth's population, the indirect effects of nuclear war would have a more serious impact than the direct effects.

It has, however, been emphasized many times that there are many remaining uncertainties in the scientific analyses of the post-nuclear war world. Some of these would be very difficult to resolve, notably those related to human decision-making. Other areas of uncertainty could potentially be reduced with further research. The following list summarizes some of the most important areas:

- *Fuel loading in and adjacent to likely targets of nuclear weapons*: Only preliminary estimates have been made of the fuel in cities, the biomass in wildlands and the worldwide production and consumption of flammable materials and fuels. More information is needed on fossil fuel inventories; the amount of combustible fuel stored in cities and in rural areas, particularly those that are in the vicinity of potential nuclear targets; and inventories of dangerous and toxic chemical stores in potential blast areas.

- *Smoke production and properties*: More information is needed on the production and properties of smoke from massive fires. Large-scale experimental fires could provide further data on the quantity of smoke produced, the fraction of soot produced, the shape, size and optical properties of the smoke particles, and other essential properties. Laboratory studies of a number of essential properties of smoke particles are also needed. It is also suggested that researchers explore the feasibility of using mobile instruments to measure smoke characteristics of unplanned urban and wildland fires.

- *Fire modelling*: Improvements are needed in the models used to study the development and spread of large-scale fires; the formation of rising fire plumes and the development of thunderclouds; the interaction and merging of plumes in close proximity; and the processes that lead to removal of particles from fire-induced clouds by precipitation. In addition, meteorological and other measurements should be made during real fires to test the validity of the models. There is also a great need to improve modelling of the development of fire plumes and fire-induced clouds, and the occurrence of precipitation removal of particles, over a middle geographical scale, before the clouds and smoke particles are distributed over a global scale.

- *Climate modelling*: Computer models have already been modified to some extent to investigate what might happen to the climate if large smoke clouds were lofted into the atmosphere. A number of further improvements are needed, such as the treatment of the removal of smoke by clouds and precipitation; investigation of how sensitive the model results are to assumptions about the season in which smoke is injected into the atmosphere; extension of simulations to periods of a year or more; and investigation of the impact on the model results of the assumption that smoke clouds would be patchy. The models have not been used to analyze the effect of a major nuclear war over periods of years to decades. This is an extremely difficult problem, but areas that should be studied include interactions between the oceans and the atmosphere over months to decades and the potential effects of massive chemical releases during a nuclear war.

- *Chemistry in the atmosphere*: Further research is needed on the emission of toxic gases from burning fossil-fuel derived products, such as plastics; the possible change in tropospheric chemistry caused by chemicals released by nuclear explosions and fires; the response of soot particles to ultraviolet gases such as ozone; and the potential build-up of air pollutants and toxic substances in river valleys, lowlands and sheltered areas.

- *Radioactivity*: There is a need for further calculations of local fallout so that overlap of radioactivity from adjacent surface bursts is more realistically treated. Further research is needed to evaluate the contribution of internal doses to the radiation damage that could occur after a nuclear war; to improve understanding of the radiation doses that might result from the potential targeting of nuclear power plants and fuel cycle facilities; and to extend calculations of global fallout to include the doses that might result from material injected into the stratosphere.

- *EMP*: There is a need for further evaluation of the impact of EMP on communications and electronic systems in the midst of an international crisis.

- Biological and physical scientists should work cooperatively in developing regional and global models that reflect the climatic consequences of nuclear war, with particular attention to feedback from the biological analysts. This would help to produce the types of information needed for ecological considerations.

- There is a need for models of environmental and ecosystem responses extending into the chronic post-nuclear war phase. These should include better estimates of chronic phase parameters of temperature, light and precipitation. These should also include much more experimental work on the effects of beta-radiation on plants and crops. Microcosm or enclosure experiments would be appropriate.

- Experiments are needed to give a better understanding of the importance and role of seed and seedling banks in world ecosystems and their vulnerability to climatic perturbations.

- Explicit experimentation is needed to investigate synergisms including, for example, the interactive effects on biota of radiation, UV-B and air pollution.

- Finally, it is recommended that the International Council of Scientific Unions should monitor and report on the status of continuing research in this field and that, due to the interdisciplinary nature of this problem, further collaboration between physical and biological scientists be encouraged.

Further Reading

Alexandrov, V. V. and Stenchikov, G. L. |1984| On global consequences of nuclear war. *J. of Computing Mathematics and Mathematical Physics |Moscow|* **24**, No. 1, 140–143.

British Medical Association |1983| *The Medical Effects of Nuclear War.* John Wiley and Sons, Chichester, UK.

Chazov, Y. I., Ilyin, L.A. and Guskova, A. K. |1984| *Nuclear War: The Medical and Biological Consequences,* Novosti Press Agency Publishing House, Moscow, 239 pp.

Cotton, W. R. |1985| Atmospheric convection and nuclear winter. *American Scientist* **73**, 275–280.

Covey C., Schneider, S. H. and Thomson, S. L. |1984| Global atmospheric effects of massive smoke injections from a nuclear war: results from general circulation model simulations. *Nature* **308**, 21–25.

Crutzen, P. J. and Birks, J. W. |1982| The atmosphere after a nuclear war: twilight at noon. *Ambio* **11**, 114–125.

Crutzen, P. J., Galbally, I. E. and Bruhl, C. |1984| Atmospheric effects from post-nuclear fires, *Climatic Change* **6**, 323–364.

Ehrlich, P. R., Harte, J., Harwell, M. A., Raven, P. H., Sagan, C., Woodwell, G. M., Berry, J., Ayensu, E. S., Ehrlich, A. H., Eisner, T., Gould, S. J., Grover, H. D., Herrera, P., Mooney, H. A., Myers, N., Pimental, D. and Teal, J. M. |1983| Long-term biological consequences of nuclear war. *Science* **222**, 1293–1300.

Ginsberg, A. S., Golitsyn, G. S. and Vasiliev, A. A. |1985| Global consequences of a nuclear war: a review of recent Soviet studies. *SIPRI Yearbook* [Stockholm International Peace Research Institute], Taylor and Francis Publishers, London and Philadelphia, pp. 107–129.

Glasstone, S. and Dolan, P. J. |eds| |1977| *The Effects of Nuclear Weapons |3rd ed.|.* U.S. Government Printing Office, Washington, D.C., U.S.A., 653 pp.

Harwell, M. A. |1984| *Nuclear Winter: The Human and Environmental Consequences of Nuclear War.* Springer Verlag.

Holdgate, M. W. and Woodman, M. J. |eds| |1978| *The Breakdown and Restoration of Ecosystems.* NATO Conference Series I: Ecology, Plenum Press, New York, U.S.A.

Izrael, Yu. A. |1983| Ecological consequences of possible nuclear war. *Meteorology and Hydrology* [Moscow] No. 10, 5–11.

National Research Council |1975| *Long-Term Worldwide Effects of Multiple Nuclear-Weapons Detonations.* National Academy of Sciences, 2101 Constitution Ave., N.W., Washington, D.C., U.S.A., 213 pp.

National Research Council |1984| *The Effects on the Atmosphere of a Major Nuclear Exchange.* National Academy of Sciences, 2101 Constitution Ave., N.W., Washington, D.C., U.S.A.

Obukhov, A. M. and Golitsyn, G. S. |1983| Possible atmospheric consequences of a nuclear conflict. *The Earth and the Universe,* No. 6, 4–7.

Royal Society of Canada |1985| *Nuclear Winter and Associated Effects: A Canadian Appraisal of the Environmental Impact of Nuclear War.* Roy. Soc. of Canada, 344 Wellington St., Ottawa, Canada.

SCOPE 28 |1985| *Environmental Consequences of Nuclear War* Vol. I: Physical and Atmospheric Effects |B. A. Pittock, T. Ackerman, P.J. Crutzen. M. McCracken, C. Shapiro and R. P. Turco, eds.| Vol II: Ecological and Agricultural Effects |M. A. Harwell and T. C. Hutchinson, eds.| John Wiley and Sons Ltd.. Chichester, UK.

Thompson, S. L., Alexandrov, V. V., Stenchikov, G. L., Schneider, S. H., Covey, C., and Chervin, R. M. |1984| Global climatic consequences of nuclear war: simulations with three dimensional models. *Ambio* **13**, 236–243.

Turco, O.R., Toon. P. B.. Ackerman, T. P., Pollack, T. P., and Sagan, C. |1983| Global atmospheric consequences of nuclear war. *Science* **222**, 1283–1292.

WHO |1983| *Effects of Nuclear War on Health and Health Services.* Rep. No. A36/12, Geneva, World Health Org.

— |1982| *Ambio* Vol II. Issue No. 2. This issue contains a basic set of papers on the environmental consequences of nuclear war. It is a departure point for a "historical" perspective on this subject.

— |1982| *Common Security.* The Report of the Independent Commission on Disarmament and Security Issues |Olaf Palme, Chairman| Pan Books, London and Sydney, 202 pp.

— |1983| National scientists' conference to save the world from the threat of nuclear war and to assure disarmament and peace. *Vestiuk*, U.S.S.R. Academy of Sciences, No. 9, pp 3–124.

— *Priroda* |1985| Issue No. 6. This issue contains several papers on nuclear winter: Editorial: Fight against war before it begins, 3–5. Koozin, A. M.: Consequences of nuclear war—views of a radiobiologist, 17–21. Golitsyn, G. C.: Atmospheric consequences of nuclear war, 22–29. Budyko, M. I.: Aerosol climatic catastrophes, 30–38. Stenchikov, G. L.: Mathematical modelling of climate. Svirezhev, Yu. M.: Long-term consequences of nuclear war—global ecological catastrophe.